Application of the Commission's Recommendations for the Protection of People in Emergency Exposure Situations

ICRP PUBLICATION 109

Approved by the Commission in October 2008

Abstract–This report was prepared to provide advice on the application of the Commission's 2007 Recommendations. The advice includes the preparedness for, and response to, all radiation emergency exposure situations defined as: 'situations that may occur during the operation of a planned situation, or from a malicious act, or from any other unexpected situation and require urgent action in order to avoid or reduce undesirable consequences'. An emergency exposure situation may evolve, in time, into an existing exposure situation. The Commission's advice for these types of situation is published in two complementary documents (that for emergency exposure situations in this report, that for existing exposure situations following emergency exposure situations in a forthcoming report entitled 'Application of the Commission's recommendations to the protection of individuals living in long-term contaminated territories after a nuclear accident or a radiation emergency').

The Commission's 2007 Recommendations re-state its principles of justification and optimisation, and the requirement to protect against severe deterministic injury, as applying to emergency exposure situations. For the purpose of protection, reference levels for emergency exposure situations should be set in the band of 20–100 mSv effective dose (acute or per year). The reference level represents the level of residual dose or risk above which it is generally judged to be inappropriate to plan to allow exposures to occur. The Commission considers that a dose rising towards 100 mSv will almost always justify protective measures. Protection against all exposures, above or below the reference level, should be optimised.

More complete protection is offered by simultaneously considering all exposure pathways and all relevant protection options when deciding on the optimum course of action in the context of an overall protection strategy. Such an overall protection strategy must be justified, resulting in more good than harm. In order to optimise an

overall strategy, it is necessary to identify the dominant exposure pathways, the time scales over which components of the dose will be received, and the potential effectiveness of individual protective options. If, in application of an overall protection strategy, protection measures do not achieve their planned residual dose objectives, or worse, result in exposures exceeding reference levels defined at the planning stage, a re-assessment of the situation is warranted. In planning and in the event of an emergency, decisions to terminate protective measures should have due regard for the appropriate reference level.

The change from an emergency exposure situation to an existing exposure situation will be based on a decision by the authority responsible for the overall response. This transition may happen at any time during an emergency exposure situation, and may take place at different geographical locations at different times. The transfer should be undertaken in a co-ordinated and fully transparent manner, and should be understood by all parties involved.

Keywords: Emergency exposure situation; Reference level; Constrained optimisation; Protection strategy

ICRP Publication 109

Editorial

LOOKING BACK, LOOKING FORWARD

The players change, perhaps the plot changes direction slightly, but the play does go on. Effective January 1st, 2009, I had the great honour of being named the Scientific Secretary of the International Commission on Radiological Protection (ICRP), the fifth since the position became full-time in 1962 with the appointment of David Sowby (another Canadian). Dr. Jack Valentin, my immediate predecessor and dear friend for many years, has been an outstanding mentor during my first months with ICRP. I must take this occasion to thank him publicly and wholeheartedly for his dedication to ICRP over many years, and his patience with me throughout this transition for both of us.

2009 also brings in a new Chair for ICRP, Dr. Claire Cousins, the 12th since Rolf Sievert first held the post in 1928. Dr. Lars-Erik Holm is retiring from ICRP, having served as ICRP Chair for 4 years, on the ICRP Main Commission for 12 years, and 8 years before that on Committee 1. At the same time, Dr. Abel González takes on the role of Vice-chair for the 2009–2013 term. New appointees to the ICRP Main Commission for this term are: Professor Eliseo Vañó (Committee 3 Chair), Jacques Lochard (Committee 4 Chair), Dr. John Cooper, and Dr. Ohtsura Niwa. In addition, there are many new Committee and Task Group members in this term.

I had the pleasure of being present at the October 2008 meeting of the ICRP Main Commission, the last of that term. It was at this meeting that Dr. Cousins was elected ICRP Chair, and I was formally introduced to the Main Commission members. It also marked the 80th anniversary of ICRP. This moment was celebrated with a toast and a small token given to Main Commission members. This token was a memory stick containing all the recommendations of ICRP and its predecessor organisation, the International X-ray and Radium Protection Commission (IXRPC): 80 years of radiological protection recommendations squeezed on to a device small enough to lose in one's pocket!

When one looks back at these recommendations, it is interesting to see how much has changed, and at the same time how little. The 1928 Recommendations (IXRPC, 1929) filled 12 pages, and were issued in English, German, and French (four pages each). They included advice about having good natural lighting and fresh air in X-ray rooms, and decorating them in light colours. Although this sounds pleasant, it is not the type of advice considered to be within the realm of radiological protection today. However, it is not difficult to recognise the still-fundamental advice on limiting work times, increasing the distance from the source, and the use of shielding.

On the subject of looking back, in this issue of the *Annals of the ICRP*, you will find an excellent article on the history of ICRP written by former ICRP Chair Dr. Roger Clarke and ICRP Scientific Secretary Emeritus Dr. Jack Valentin. Although this article is not part of the recommendations of ICRP, I hope you will find it both interesting and enlightening.

The main subject of this issue is the ICRP recommendations on emergency exposure situations. These provide advice on the application of the Commission's 2007 Recommendations (ICRP, 2007) with respect to preparedness for and response to all emergency exposure situations. This is an area in which the Commission's 2007 Recommendations have evolved in some significant ways from those of 1990 (ICRP, 1991). For example, rather than assessing the potential benefits of individual protection options, the approach is now to consider all exposure pathways and all relevant protection options when deciding on the optimum course of action in the context of an overall protection strategy. The objective is the best possible overall response under the circumstances; something that was difficult to ensure when considering individual actions in isolation.

Once immediate actions have been taken and the situation has stabilised, under some circumstances, one will be left with the very different problem of residual contamination. Within the ICRP system of protection, this represents an evolution from an emergency exposure situation to an existing exposure situation. Actions are no longer truly urgent, and therefore a more measured approach can be taken to deal with the remaining problems. Many of the difficulties that may be encountered in this transition are discussed within this report.

Looking forward, a report on 'Application of the Commission's recommendations to the protection of individuals living in long-term contaminated territories after a nuclear accident or a radiation emergency' will be published in the not-too-distant future. In a sense, with respect to post-emergency situations, it will pick up where the current report leaves off. The Task Groups working on these two documents have co-ordinated their efforts so that they give complementary advice of use to radiological protection professionals in the field of emergency and consequence management.

It is a privilege to serve as the Editor of this august journal, and it has been a pleasure to write this editorial. However, this is only one of the many roles played by the Scientific Secretary. Many aspects of the office are daunting to say the least, but heightened challenges can bring with them a greater sense of satisfaction when the job is well done. To excel, I know that a balance must be found between following the script written by those before me and making the role my own. That said, dear friends, colleagues, and readers at large, this is not a role I take on alone. I know that I can count on you to support the work of ICRP, in particular through continuing to devote your time, energy, and experience to providing constructive feedback on ICRP publications and the work of ICRP in general. The end result can only be improved recommendations and a more widespread and in-depth understanding of the system of protection, ultimately making this stage upon which we all act just a little safer.

Christopher H. Clement
Scientific Secretary, ICRP

References

ICRP, 1991. 1990 Recommendations of the International Commission on Radiological Protection. ICRP Publication 60. Ann. ICRP 21(1–3).

ICRP, 2007. The 2007 Recommendations of the International Commission on Radiological Protection. ICRP Publication 103. Ann. ICRP 37(2–4).

IXRPC, 1929. International Recommendations for X-ray and Radium Protection. Stockholm. P.A. Norstedt & Söner.

CONTENTS

PREFACE

Between 31 October and 3 November 2006, the Main Commission of the International Commission on Radiological Protection (ICRP) approved the formation of a new Task Group, reporting to Committee 4, to develop guidance on the implementation of the new ICRP recommendations on the principle of the optimisation of radiological protection in various states of emergency preparedness and response.

As stated in the terms of reference, the objective of the Task Group was to develop a report on the application of the Commission's recommendations for the protection of populations in the emergency phase of a nuclear accident or a radiological emergency. It was tasked, in particular, to provide guidance on:

- the setting of reference levels in both planning and implementation of an emergency response;
- how reference levels assist emergency response management;
- how optimisation can be applied in identifying protective measures at the planning stage;
- the management of changing protective measures with time; and
- the interface with the rehabilitation phase.

Particular attention has been given to the interface with the rehabilitation phase following a nuclear accident or a radiological emergency through a close co-ordination with the Task Group developing recommendations on this aspect.

This report takes into account recent developments, views, and experiences in emergency management. Ongoing work and efforts performed by international organisations, e.g. the revision of the International Basic Safety Standards, have also been considered. The guidance offered by the Task Group is generic, providing a basic framework that can be tailored to specific circumstances. The Task Group considers that detailed implementation of the Commission's recommendations is a matter for the relevant national authorities.

The guidance in this report builds upon the concept of optimisation of protection below reference levels as recommended previously by ICRP.

The membership of the Task Group was as follows:

W. Weiss (Chairman)	J. Fairobent	M. Morrey
O. Pavlovsky	D. Queniart	

The corresponding members of this Task Group were:

E. Buglova	T. Lazo	I. Robinson

The membership of Committee 4 during the period of preparation of the report was:

A. Sugier (Chairperson)	P.A. Burns	P. Carboneras Martinez
D. Cool	J.R. Cooper (Vice-Chair)	J.-F. Lecomte (Secretary)
H. Liu	J. Lochard	G. Massera
A. McGarry	M. Kai	K. Mrabit
M. Savkin (–2008)	K.-L. Sjöblom	A. Simanga Tsela
W. Weiss		

The Task Group would like to thank the organisations and staff that made facilities and support available for its meetings. These include the Federal Office for Radiation Protection in Germany and the Nuclear Energy Agency of the Organisation for Economic Co-operation and Development.

The report was adopted by the Commission at its meeting in Buenos Aires on 25 October 2008.

The recommendations and guidance provided by the Commission are usually attributed to the Commission itself, i.e. the reports represent a formal statement agreed by the entire Commission. Such reports are thus, technically, anonymous, although the membership of the Task Group drafting a report is always stated in the preface.

On rare occasions, reports have included reviews or appendices written by, and attributed explicitly to, individual members of the Commission or one of its Committees or Task Groups. Such contributions should be understood to have the same status as invited or proffered papers in any peer-reviewed journal, i.e. the Commission and ultimately the Editor (the Scientific Secretary of ICRP) regards the contribution as worthy of publication, but does not necessarily agree with its contents, which are the responsibility of the named author(s) alone.

The Commission feels that such contributions can make the *Annals of the ICRP* more useful, and intends to include such material somewhat more frequently in the future. At its meeting in Buenos Aires in October 2008, the 80th anniversary of ICRP, the Commission decided to invite a review on the historical development of ICRP recommendations, based on R.H. Clarke's presentation concerning the topic at the XII Congress of the International Radiation Protection Association in 2008.

The authors of this review are:

Dr. Roger Clarke Dr. Jack Valentin

EXECUTIVE SUMMARY

Basic principles

(a) The Commission's 2007 Recommendations (ICRP, 2007) re-state its principles of justification and optimisation as applying to emergency exposure situations. This means that the level of protection should be the best possible under the prevailing circumstances, maximising the margin of benefit over harm. In order to avoid grossly inequitable outcomes of the optimisation procedure, the process should be constrained, to the extent practicable, by restrictions on the dose or risks received by individuals as a result of the emergency.

(b) The reference level represents the level of residual dose or risk above which it is generally judged to be inappropriate to plan to allow exposures to occur. Therefore, any planned protection strategy should at least aim to reduce exposures below this level, with optimisation achieving still lower exposures. Protection against all exposures, above or below the reference level, should be optimised. In the context of developing response plans for emergency exposure situations, the Commission recommends that national authorities should set reference levels between 20 mSv and 100 mSv effective dose (acute or per year, as applicable to the emergency exposure situation under consideration). Reference levels below 20 mSv may be appropriate for the response to situations involving low projected exposures. There may also be situations where it is not possible to plan to keep all doses below the appropriate reference level, e.g. extreme malicious events or low-probability, high-consequence accidents in which extremely high acute doses can be received within minutes or hours. For these situations, it is not possible to plan to avoid such exposures entirely. Therefore, the Commission advises that measures should be taken to reduce the probability of their occurrence, and response plans should be developed that can mitigate the health consequences where practicable.

(c) The Commission now considers that more complete protection is offered by simultaneously considering all exposure pathways and all relevant protection options when deciding on the optimum course of action. While each individual protective measure must be justified by itself in the context of an overall protection strategy, the full protection strategy must also be justified, resulting in more good than harm. This approach may represent a relative increase in operational complexity, but it also provides a significant amount of increased flexibility in designing the optimum protection to address an emergency exposure situation by focusing on the combined effects of all individual protective measures included in the protection strategy rather than on any single protective measure. Moreover, the new approach provides a framework that supports a consideration of how individual protective measures affect one another, and it helps focus resource allocation to where the strongest overall benefit can be achieved. It also recognises that the dose which an individual has already received during an emergency should be taken into consideration when determining what constitutes optimum protection in later response actions.

(d) In order to optimise an overall planned protection strategy, it is necessary to identify the dominant exposure pathways, the time scales over which components of the dose will be received, and the potential effectiveness of individual protective options. Knowledge of the dominant exposure pathways will guide decisions on the allocation of resources. Resource allocation should be commensurate with the expected benefits, of which averted dose is an important component. Knowledge of the time periods over which exposures will be received informs decisions about the lead times available to organise the implementation of protective measures once an emergency exposure situation has been recognised. Where urgent actions are required to reduce exposures, specific legislation would facilitate efficient management of the response (e.g. management of contaminated wastes). Furthermore, it is important to use easily identifiable 'triggers' as the basis for decisions to implement urgent protective measures.

(e) The Commission recognises that there is a qualitative difference between the risks of stochastic health effects and the risks of an individual receiving exposures that would result in severe deterministic injury. By 'severe deterministic injury', the Commission means injuries that are directly attributable to the radiation exposure, irreversible in nature, and severely impair the quality of life of individuals, e.g. lung morbidity and early death. The Commission recommends that every practicable effort should be made to avoid the occurrence of severe deterministic injuries in an emergency exposure situation. This means that it will be justified to expend significant resources, both at the planning stage and during the response, if this is required, in order to reduce expected exposures to below the thresholds for these effects. Furthermore, where prompt medical intervention has the potential to avert such injury, the Commission recommends that procedures and measures should be included in the emergency response plan to enable those individuals who may have received such high exposures to be identified promptly and receive appropriate medical treatment.

Arrangements for emergency exposure situations

(f) The Commission recommends that plans should be prepared for all types of emergency exposure situation: nuclear accidents (occurring within the country and abroad), transport accidents, accidents involving sources from industry and hospitals, malicious uses of radioactive materials, and other events, such as a potential satellite crash. The level of detail within a plan will depend on the level of threat posed, and the degree to which the circumstances of the emergency can be determined in advance. However, even outline generic plans should indicate the responsibilities of different agencies, methods for communication and organisation between them during the response, and a framework for guiding decision making. More detailed plans should contain a description of the overall protection strategy, and provide triggers for initiating those aspects of the response that need to be implemented promptly. It is for the relevant national authorities to determine the detail of planning that is appropriate for different situations.

(g) It is essential that all aspects of the plan are consulted with relevant stakeholders, otherwise it will be more difficult to implement them during the response. To the extent possible, the overall protection strategy and its constituent individual protective measures should be worked through and agreed with all those potentially exposed or affected. Such an engagement will assist the emergency plans in being focused on the protection of those at greatest risk in the initial phases, and also on the progression to populations resuming 'normal' lifestyles.

(h) In the event of an emergency exposure situation, it is likely that exposure rates will vary in space and time, and that the doses received by individuals will vary, both as a result of the variations in exposure rates and as a result of differences in their physiological characteristics and behaviours. These population groups should be characterised by representative persons, as described in the Commission's advice on the representative person. In accordance with the Commission's advice on the representative person, it is important that the dose estimates reflect those likely to be received by the groups at greatest risk, but that they are not grossly pessimistic.

(i) The Commission's band of reference levels is expressed in terms of effective dose. For many emergency plans, this is an appropriate quantity in which to express the reference level. However, there are situations for which effective dose is not an appropriate quantity to express reference levels. This is the case when the type or scale of an emergency may result in doses in excess of 100 mSv effective dose (where the assumption of linearity may no longer hold), when parts of the response need to focus on individuals at risk of incurring severe deterministic injury, and when the resulting exposures are very strongly dominated by irradiation of a single organ for which very specific protective measures are optimum (e.g. releases dominated by radioiodine). For these situations, the Commission advises that consideration should be given to specifying (or providing supplementary) reference levels in terms of organ dose.

(j) In its previous advice, the Commission recommended the use of intervention levels of averted dose to assist decisions on whether/when to include certain protective measures in an overall protection strategy. It should be emphasised that the intervention level is understood as a level above which an action is justified and below which no action (e.g. no optimisation of protection) is needed. This concept is no longer valid. Moreover, the Commission now recommends focusing on optimisation of protection with respect to the overall protection strategy, which includes exposures from all exposure pathways simultaneously, rather than individual measures. However, the levels of averted dose recommended in *Publication 63* (ICRP, 1991a,b) for optimisation of protection in terms of individual protective measures may still be useful as inputs to the development of the overall response (see also ICRP, 2005).

(k) In order to develop an emergency plan, it is necessary to evaluate the projected doses for the situations being considered. The purpose of estimating projected doses and their likely spatial and temporal distributions is three fold: first to identify the scale of health consequences that might occur if no protective measures were taken (and, in particular, whether there is a risk of severe deterministic injury), and from this to determine the broad scale of resources it is appropriate to assign to a protection strategy; second, to identify the broad geographical and temporal distribution of the various likely response phases; and third, where, in terms of protection, resources are likely to be spent most effectively. Where it is judged appropriate to develop a detailed emergency response plan, it is important to identify whether specific provisions are required to protect those at risk of severe deterministic injury. If so, this part of the plan should be given priority for focus and resources, and should be separately justified and optimised.

(l) For detailed planning to protect against exposures resulting in stochastic risk, it is useful to begin the development of an overall protection strategy by identifying all protective measures that are likely to be justified, even if they only avert a relatively small component of the projected dose. Once all protective measures that are likely to be individually justified have been identified, each one should be examined for its potential to avert a significant proportion of the projected dose, and for consequences that may interact with those of other protective measures in such a way as to render their combined implementation either significantly more strongly justified or unjustified. From this initial scoping review, a broad outline protection strategy can be developed.

(m) Having identified protective measures that are likely to be included in the protection strategy, it is necessary to evaluate the residual doses (i.e. those to different representative individuals) that would result from implementing the protection strategy. The first step is to scope the residual doses, in order to compare them with the appropriate reference level. If the residual dose is likely to be below the reference level, detailed optimisation of the protection strategy can be undertaken. If not, changes to the protective measures or their implementation need to be considered, and the process of comparison of the reference level with the residual dose repeated.

(n) Some combinations of protective measures can be considered to be largely independent of one another, e.g. commercial food restrictions and the evacuation of populations in close proximity to a radiation source. These types of protective measures can readily be optimised separately and their relevant averted doses can be used as a direct guide.

(o) The resources required to implement protective measures are not the only factors that might interact within an overall protection strategy. Other such factors include individual and social disruption, anxiety and reassurance, and indirect economic consequences. It is important to review the proposed overall protection strategy with relevant stakeholders to ensure that the plan is optimised with respect to these factors, as well as with respect to dose and the resources required. This wider review of the protection strategy may indicate a role for additional measures which, in isolation, may not appear optimum (or even justified).

(p) Once the protection strategy has been optimised, triggers for initiating the different parts of an emergency response plan for the early phase should be developed. Triggers may be expressed in terms of any observable circumstances or directly measurable quantities, such as plant conditions, dose rates, or wind direction. They may be related to dose considerations, but are more likely to be related to key indicators of the occurrence of the emergency situation for which the plan (or a group of protective measures within the plan) was developed. It may not be appropriate to specify triggers for initiating protective measures later in the plan, since these should generally take account of the specific details of the evolving emergency situation. For such protective measures, it may be helpful to include in the response plan an agreed framework for developing triggers in 'real time' when needed. The inclusion of such a framework is likely to result in wider acceptance of the 'real-time' triggers when they are developed.

Implementing protection strategies

(q) In the context of the ICRP system of radiological protection, there is one fundamental difference between prospectively planning to address the consequences of a radiological emergency exposure situation, and managing consequences that are in the process of occurring or that have already occurred. In the context of planning, optimisation is performed using the appropriate reference level as the upper bound, eliminating all protection solutions that result in individual residual doses exceeding the reference level. The inherently unpredictable nature of emergency exposure situations, their tendency to evolve rapidly, and the wide possible range of emergency conditions (i.e. weather conditions, geographical location, population habits, etc.) could result in situations that do not match the assumptions that were used to develop the optimised protection strategies, and some actual exposures may exceed the preselected reference level. As such, in the context of managing the consequences of an emergency that is in the process of occurring or that has already occurred, the predefined reference level is used as a benchmark against which the results of implementing an optimised protection strategy can be judged, and for guiding the development and implementation of further protective measures if necessary.

(r) Once an emergency exposure situation has occurred, it is likely that many stakeholders will be very interested in providing input to discussions regarding protective measures. Should the emergency exposure situation require urgent protective measures, the 'reflex' use of preplanned protection strategies, implemented on the basis of predefined triggers, will be necessary with no or very little stakeholder involvement beyond the emergency response authorities and those responsible for the site, facility, or source that is causing the emergency exposure situation. Inappropriate involvement of stakeholders or excessive review of the detailed effectiveness of such 'reflex' protective actions is likely to reduce their effectiveness by delaying their implementation, and this should be avoided. However, as the emergency exposure situation progresses, it will become increasingly beneficial to involve stakeholders in discussions leading to protection decisions. It is therefore important that part of emergency response planning should be the development and implementation of processes and procedures to inform and involve stakeholders once the most urgent protective actions have been implemented.

(s) In many cases, emergency response planning will broadly fit a large range of possible situations, such that the timely implementation of a planned protection strategy should come close to providing the optimum protection, with divergence most likely being on the conservative side. However, there may be a need to operationally adjust planned protection strategies, justifying new protective measures or significant changes to plans. The need to consider such modifications may increase as the emergency exposure situation progresses, and the magnitude of changes from plans may depend on the nature of the emergency exposure situation that occurs.

(t) If, in application, protective measures do not achieve their planned residual dose objectives, or worse, result in exposures exceeding the reference levels defined at the planning stage, a re-assessment of the situation is warranted to understand why plans and results differ so significantly. New protective measures could then, if appropriate, be selected, justified, optimised, and applied, or existing options could be extended in time and/or space.

(u) As an emergency exposure situation progresses and understanding of the exact circumstances increases, decisions will increasingly be based on actual circumstances rather than on preplanned responses, assumptions, and models. There will also be an increased need to plan future protection strategies in greater detail than included in the initial emergency plan.

(v) The decision to terminate individual protective measures will need to reflect the prevailing circumstances of the emergency exposure situation being addressed in an appropriate manner. For the termination of early protective measures, guidance should have been developed and included in the emergency plan. For later protective measures, wherever possible, the criteria for terminating the measures should be agreed with relevant stakeholders in advance of their implementation. In this case, criteria for termination are best expressed in terms of directly observable or measureable quantities, so that achievement of the criteria can be demonstrated clearly. In planning and in the event of an emergency, decisions to terminate protective measures should have due regard for the appropriate reference level. In planning, this is an integral part of the optimisation of the protection strategy. However, since the actual circumstances of an emergency may deviate from those addressed during planning, it is important to consider the implications for residual dose when making decisions regarding the termination of protective actions, using the reference level as a benchmark.

Transition to rehabilitation

(w) The Commission recommends that the management of exposures in the long term following an emergency exposure situation should be treated as an existing exposure. This is because the characteristics of response become very different from those at an early stage. The management of existing exposure situations involves accepting that the exposure situation is different from what would normally be considered acceptable, but recognising that, given the circumstances and possibly subject to some ongoing special measures, the exposure can and will be tolerated, i.e. that stability has been achieved.

(x) The change from an emergency exposure situation to an existing exposure situation will be based on a decision by the authority responsible for the overall response. This transition may happen at any time during an emergency exposure situation, although not generally when urgent actions are being taken. Moreover, this transition may take place at different geographical locations at different times, such that some areas are managed as an emergency exposure situation whilst others are managed as an existing exposure situation. The transition may require a transfer of responsibilities to different authorities. This transfer should be undertaken in a co-ordinated and fully transparent manner, and should be understood by all parties involved. The Commission recommends that planning for the transition from an emergency exposure situation to an existing exposure situation should be undertaken as part of the overall emergency preparedness, and should involve relevant stakeholders.

(y) Existing exposure situations which are created by emergency exposure situations can be characterised as having some sort of residual exposure pathways and lingering contamination above previous background levels, but having social, political, economic, and environmental aspects of the situation that will be sustained, and are seen by the affected populations and governments as being their new reality. There are no predetermined temporal or geographical boundaries that delineate the transition from an emergency exposure situation to an existing exposure situation. In general, a reference level of the magnitude used in emergency exposure situations will not be acceptable as a long-term benchmark, as these exposure levels are generally unsustainable from social and political standpoints. As such, governments and/or regulatory authorities will, at some point, have to identify and set a new reference level, typically at the lower end of the range recommended by the Commission of between 1 and 20 mSv/year.

(z) For some large emergency situations involving the release of high levels of long-lived contamination over large areas, part of the new reality following the situation may be that some areas will be so contaminated as to be incapable of sustaining social, economic, and political inhabitation as before. In these areas, governments may prohibit human habitation and other land uses. This would mean that any populations evacuated from these areas would not be allowed to return, and that further resettlement or use of these areas would not be allowed. Clearly, it is not easy for a government and its people to make a decision to remove people permanently (or for the long-foreseeable future) from an area and to forbid its use. As such, the social, economic, political, and radiological aspects of such a choice will need to be discussed in a broad and transparent fashion before a decision is reached.

References

ICRP, 1991a. 1990 Recommendations of the International Commission on Radiological Protection. ICRP Publication 60. Ann. ICRP 21(1–3).

ICRP, 1991b. Principles for intervention for protection of the public in a radiological emergency. ICRP Publication 63. Ann ICRP 22(4).

ICRP, 2005. Protecting people against radiation exposure in the event of a radiological attack. ICRP Publication 96. Ann. ICRP 35(1).

ICRP, 2007. The 2007 Recommendations of the International Commission on Radiological Protection. ICRP Publication 103. Ann. ICRP 37(2–4).

1. INTRODUCTION

(1) The Commission has set out general principles for planning an intervention in the case of a radiation emergency (ICRP, 1991a,b), and additional relevant guidance (ICRP, 2005a,b,c). More recently, the Commission has published new recommendations relating to its overall system of protection (ICRP, 2007a). The 2007 Recommendations are intended to complement, rather than replace, the Commission's previous advice. However, the advice contained in the 2007 Recommendations may have implications for emergency preparedness and response. This report discusses the application of the new advice, and explains how the previous advice fits into the revised overall system of protection. Where the Commission's advice is unchanged from its previous recommendations, or issues are discussed thoroughly in publications by other international organisations, appropriate references are given and no detailed discussion is provided. The report does not cover emergency situations involving unintended exposures of patients; these situations are dealt with separately by the Commission (ICRP, 2007b).

1.1. References

ICRP, 1991a. 1990 Recommendations of the International Commission on Radiological Protection. ICRP Publication 60. Ann. ICRP 21(1–3).

ICRP, 1991b. Principles for intervention for protection of the public in a radiological emergency. ICRP Publication 63. Ann. ICRP 22(4).

ICRP, 2005a. Protecting people against radiation exposure in the event of a radiological attack. ICRP Publication 96. Ann. ICRP 35(1).

ICRP, 2005b. Prevention of high-dose-rate brachytherapy accidents. ICRP Publication 97. Ann. ICRP 35(2).

ICRP, 2005c. Radiation safety aspects of brachytherapy for prostate cancer using permanently implanted sources. ICRP Publication 98. Ann. ICRP 35(3).

ICRP, 2007a. The 2007 Recommendations of the International Commission on Radiological Protection. ICRP Publication 103. Ann. ICRP 37(2–4).

ICRP, 2007b. Radiological protection in medicine. ICRP Publication 105. Ann. ICRP 37(6).

2. SCOPE OF THIS ADVICE

(2) This advice relates to preparedness for, and response to, all radiation emergency exposure situations. The Commission defines radiation emergency exposure situations as: 'situations that may occur during the operation of a planned situation, or from a malicious act, or from any other unexpected situation and require urgent action in order to avoid or reduce undesirable consequences'. The scope of this advice is preparedness for, and response to, emergency exposure situations. It covers the protection of all those at risk of exposure, whether they are directly involved in mitigating actions (termed emergency 'workers' in this report, regardless of whether or not they are routinely exposed to radiation as a result of their normal employment), or are simply in need of protection (termed 'the public' in this report).

(3) An emergency exposure situation that includes the release of significant quantities of longer-lived radionuclides may evolve, in time, into an existing exposure situation. The management of emergency exposure situations and the management of existing exposure situations have distinct characteristics. Therefore, the Commission's detailed advice for these situations is published in two complementary documents (that for emergency exposure situations in this report, and that for existing exposure situations following emergency exposure situations in a forthcoming ICRP publication.

(4) Various design aspects of facilities where radioactive materials are present affect both the likelihood of an emergency arising, and also the magnitude of doses in the event that an emergency occurs. Such design measures, including both passive and active safety features, should be considered in a prior safety assessment that considers all reasonably foreseeable events. Such assessments are outside the scope of this report.

3. OBJECTIVES OF PROTECTION IN EMERGENCY EXPOSURE SITUATIONS

(5) In the event of an emergency exposure situation, the primary concern is the prevention or reduction of radiation dose. However, the potential consequences are wider ranging than the risk of radiation health effects. As demonstrated by the accident at the Chernobyl nuclear power station in 1986, the social and economic consequences may be serious and extend over a prolonged period of time. The goals for response must therefore encompass these wider potential impacts. A number of international bodies have summarised the practical goals of emergency response to radiation emergency as:

- Goal 1: to regain control of the situation;
- Goal 2: to prevent or mitigate consequences at the scene;
- Goal 3: to prevent the occurrence of deterministic health effects in workers and the public;
- Goal 4: to render first aid and manage the treatment of radiation injuries;
- Goal 5: to reduce, to the extent practicable, the occurrence of stochastic health effects in the population;
- Goal 6: to prevent, to the extent practicable, the occurrence of adverse non-radiological effects on individuals and among the population;
- Goal 7: to protect, to the extent practicable, the environment and property; and
- Goal 8: to take into account, to the extent practicable, the need for resumption of normal social and economic activity.

The Commission agrees broadly with these goals. This report explains how the application of the Commission's advice will contribute to their achievement.

(6) The Commission recognises that there is a qualitative difference between the risks of stochastic health effects and the risks to an individual receiving exposures that would result in a severe deterministic injury. By 'severe deterministic injury', the Commission means injuries that are directly attributable to the radiation exposure, irreversible in nature, and severely impair the quality of life of that individual, e.g. lung morbidity and early death. The Commission recommends that every practicable effort should be made to avoid the occurrence of severe deterministic injuries in an emergency exposure situation, and that planning to protect against the occurrence of severe deterministic injury should take priority over that to protect against stochastic risks.

(7) The results of an analysis of the lessons learned from the responses to the Chernobyl, Goiania, and other emergencies over the past years lead to the conclusion that while the nature and extent of past emergencies are dissimilar, the lessons concerning emergency response are very similar, see below.

- Non-experts (the public) and decision makers in different fields implement protective and other actions.
- The public and decision makers want to know that they and their loved ones are safe. A rationale based solely on cost benefit and averted dose is not helpful in addressing this concern.

- Criteria consistent with established radiation protection principles cannot be developed effectively solely during or after an emergency because communication of those criteria can become more difficult.
- Non-radiological (e.g. economic, social, and psychological) consequences may become more important than the radiological consequences due to a lack of pre-established guidance that is understandable to the public and officials, and because of the nature of actual prevailing circumstances.

(8) It is important to prepare an agreed framework within which decisions on the optimum response will be made. Such an agreed framework should represent an overall protection strategy which does not focus solely on actions required after the onset of an emergency exposure situation.

(9) For many types of emergency exposure situation, exposure rates will be greatest immediately following the event and will then decrease with time (or the extent of uncertainty surrounding exposure rates will mean that this is a prudent assumption to make for protection purposes). This means that some protective actions (e.g. sheltering, evacuation) need to be taken promptly in order to be effective. To implement these actions, there is no time to undertake detailed exposure assessments in real time. It is therefore necessary to determine, in advance, a set of internally consistent criteria for taking such actions, and, based on these criteria, to derive appropriate triggers (expressed as readily measurable quantities or observables) for initiating them in the event of an emergency.

(10) As the emergency exposure situation evolves in time, it may become prudent to extend the geographical or temporal spread of initial protective measures, and other protective measures may become appropriate, such as decontamination. Since the initial protective measures will have provided significant protection for those at greatest risk, decisions to implement other, less urgent, protective measures need to consider the actual circumstances of the situation and the optimisation of the overall protection strategy more carefully. Therefore, it will not always be appropriate to define the implementation criteria precisely in advance for less urgent protective actions. Where appropriate, the procedures by which the protective actions implemented by such criteria would be justified and optimised during an emergency should be agreed in advance, in order to facilitate their acceptance by the public during the emergency. Scientifically based recommendations for implementing protective and other measures need to be accompanied by an explanation that enables the decision maker to understand and consider them, and also to explain them to the public.

(11) The Commission recognises that the nature and extent of stakeholder involvement may vary between countries, but suggests that engagement with stakeholders is an important component in the justification and optimisation of protection strategies in emergency exposure situations. In this context, the stakeholders referred to by the Commission may include many different types of people and organisations, e.g. the public affected by the emergency exposure situation, the authority responsible for the emergency response, the licensee – if there is one – of the facility or activity causing the emergency exposure situation, the regulatory authority licensing the

facility or activity – if there is one – causing the emergency exposure situation, local public officials within and perhaps near the areas affected by the emergency exposure situation, emergency workers including first responders, and others. The stakeholders involved in any particular aspects of an emergency response situation will vary with the type of situation/facility being considered, the scale of the emergency exposure situation being considered, and the time phase of the emergency exposure situation being addressed.

4. PROTECTION OF EMERGENCY WORKERS

(12) Emergency workers and their roles should be identified in advance. Emergency workers may include radiation workers (e.g. employees of registrants and licensees) and people who are not normally occupationally exposed to ionising radiation, such as police, rescue personnel, fire fighters, and medical personnel.

(13) All workers identified in an emergency plan should have appropriate training sufficient to carry out their emergency role, so that they have sufficient information upon which to base informed consent should that be needed, and so that they can contribute to their own protection. They should also be provided with personal protective equipment, and arrangements should be made to assess any radiation doses received.

(14) The exposure of workers responding to an emergency who are implementing an emergency plan can generally be seen as deliberate and controlled, although this is not always the case; thus, some flexibility is required. Therefore, where feasible, the system of radiological protection consistent with that for planned exposure situations should be applied. Nevertheless, there may be a need to take protective actions promptly during an emergency, necessitating exposures for some workers higher than the dose limit for planned exposure situations (such as to help endangered people or to prevent the exposure of a large number of people). In such cases, it may be acceptable for emergency workers to receive, on the basis of informed consent, doses that exceed the occupational dose limits normally applied. Nevertheless, such doses should be optimised and be below a predetermined dose level appropriate to the type of task undertaken. The predetermined guidance values should take into account the assessment upon which the emergency plan is based, together with expert radiation protection advice.

(15) Previously (ICRP, 1991), the Commission advised that the exposures of emergency workers should be managed by grouping them into three categories:

- Category 1: those engaged in urgent action at the site of the accident;
- Category 2: those implementing early protective actions and taking action to protect the public; and
- Category 3: those implementing recovery operations during the intermediate phase.

Additional advice for protecting responders to a radiological attack was given in *Publication 96* (ICRP, 2005).

(16) The Commission's advice with respect to Categories 1 and 3 is essentially unchanged. With respect to Category 2, the Commission now recommends that protection should be consistent with the full system for planned exposure situations where this is feasible. This is a small change from the advice in *Publication 63* (ICRP, 1991), where the focus was on planning exposures so that they did not exceed those 'permitted in normal conditions'. The new advice can be thought of as requiring the optimisation of protection below a reference level of dose that is equivalent to the occupational exposure dose limit. Workers carrying out Category 2 tasks would

be expected to include ambulance crews, medical personnel, drivers of evacuation vehicles, and police (fire fighters and rescue personnel may carry out Category 2 tasks; however, they may also carry out Category 1 tasks).

(17) The Commission's advice, as set out in *Publication 63* (ICRP, 1991), regarding the appropriate provision of training and information to emergency workers, and, for those in Category 1, ensuring that they are undertaking the risks voluntarily, remains unchanged. Furthermore, the Commission now explicitly recommends that women who have declared that they are pregnant, or who are nursing an infant, should not have an emergency role that would be expected to lead to doses greater than 1 mSv or to significant contamination.

4.1. References

ICRP, 1991. Principles for intervention for protection of the public in a radiological emergency. ICRP Publication 63. Ann ICRP 22(4).

ICRP, 2005. Protecting people against radiation exposure in the event of a radiological attack. ICRP Publication 96. Ann. ICRP 35(1).

5. DESCRIPTION OF EMERGENCY EXPOSURE SITUATIONS

(18) *Publication 103* defines an emergency exposure situation as a situation '...that may occur during the operation of a planned situation, or from a malicious act, or from any other unexpected situation, and require urgent action in order to avoid or reduce undesirable consequences' (ICRP, 2007, Para. 176). Emergency exposure situations are characterised by the need to manage a changing situation back to one of 'normality', or at least one which is both stable and acceptable. Emergency exposure situations may be characterised by one or more of the following features: significant uncertainty concerning current and future exposures, rapidly changing rates of actual exposure, potentially very high exposures (i.e. those with the potential to cause severe deterministic health effects), or loss of control of the source of the exposure or release. Any or all of these features may continue to dominate how the response is managed for an extended period of time (i.e. months or even years), although, for some types of accident, the emergency exposure situation may be very short (days or even hours).

(19) Emergency exposure situations may be caused by many different types of initiating events and initial locations. Emergencies can occur, for example, at nuclear sites, medical facilities using radioactive materials, industrial sites that use or make radioactive sources or process materials containing naturally occurring radioactive material, or during the transport of radioactive materials, whether for commercial, energy generation or weapons use. For these situations, because the use of the radioactive material is regulated and therefore planned or known about in advance, it is possible to develop a protection strategy tailored for the specific characteristics of the potential accidents. The level of detail required for such plans will be decided by the relevant authorities, with response planning for more likely accidents developed in more detail than accidents judged to be more unlikely.

(20) Exposures could also be caused maliciously, e.g. through the dispersal of radioactive material in a public place, or arise without warning in unexpected locations, e.g. radioactive material that has by-passed regulatory controls such as 'orphan' sources. For these, it is not possible to plan a protection strategy in detail because the exact mechanism and location of exposure cannot be known in advance. However, this does not preclude the preparation of generic protection strategies with response plans; the inclusion of flexibility is of paramount importance to enable these generic plans to be adapted to the actual situation that arises. The guidance developed in this report is also applicable to these types of emergencies. Further guidance on response planning for malicious events can be found in *Publication 96* (ICRP, 2005).

(21) For planning and response to emergency exposure situations, various time-related 'phases' are often used in national response plans. The Commission recognises that different national approaches will be used with respect to various phases of an emergency, and feels that the recommendations as expressed in this report for

emergency exposure situations can be appropriately adapted to any national approach that is taken.

(22) It is necessary to define a conceptual set of doses for use in the justification and optimisation of emergency plans and decisions. These doses are as follows.

- Projected dose: dose expected to be received in the absence of the planned protective measures.
- Residual dose: projected dose minus the averted dose; dose expected to be received or measured/assessed following implementation of the planned protection strategy.
- Averted dose: dose expected to be avoided through implementation of the planned protective actions. In general, averted dose refers to the implementation of individual protective actions, but may, if specified, refer to the dose avoided from the implementation of several protective actions.

The respective roles of these doses in emergency response planning are discussed below.

5.1. Projected dose

(23) The projected dose is the individual effective (or equivalent) dose that is expected to occur as a result of an emergency exposure situation if no protective measures are employed. Projected doses should be calculated to representative persons. Generally, these will represent population groups, but, where there is a risk that individuals may be exposed above the thresholds for severe deterministic injuries, the representative persons may be assumed to undertake activities leading to the highest potential exposure. Projected doses may be used in several ways within emergency response planning:

- to give an initial indication of the scale of response planning required, by comparing them with the appropriate reference level(s);
- to determine the dominant exposure pathways and the likely time evolution of doses, for informing the emergency response planning process with respect to the type and urgency of the protective measures required; and
- to compare with threshold doses for severe deterministic injuries.

In each case, it is important that any assumptions made in the calculation of projected doses are consistent with the assumptions underlying the comparison level(s).

5.2. Residual dose

(24) The residual dose is the effective dose from all exposure pathways remaining after the implementation of an optimised protection strategy. The residual dose is the quantity that is compared with the appropriate reference level when selecting and assessing protection strategies. It can be assessed by estimating the exposure during emergency response planning (e.g. as the difference between the projected dose and the dose averted by implementing a protective measure or combination of protective

measures), or by measuring and/or calculating the actual dose after an emergency exposure situation has occurred. The residual dose should be calculated as realistically as possible.

(25) Since emergency plans are developed to protect population groups, rather than specific individuals, the residual dose is derived in planning as the dose to each of a set of representative persons. Guidance on characterising the representative person is provided in *Publication 101* (ICRP, 2006). In principle, those populations that may be exposed during an emergency should be divided into groups which are relatively homogeneous with respect to exposure and risk from that exposure, and representative persons should be characterised for each group.

(26) Once an emergency exposure situation has occurred, the residual dose should be assessed for those individuals who have been or may have been exposed. Where reasonably possible, this should be based on the assessment of exposures for real individuals. Where this is not reasonably possible, efforts should be made to characterise actual groups of exposed individuals more directly so that calculations related to representative person exposures are more accurate. As an emergency exposure situation progresses in time, more effort should be made to assess actual individual exposures.

(27) The assessment of residual dose is important when planning the response for an emergency because it is necessary to explore whether or not the dose is radiologically and socially acceptable, given the circumstances. In particular, this is fundamental to the Commission's approach to emergency response planning, and supports the achievement of Goals 3, 5, 6, 7, and 8, as listed in Section 3. The residual dose should be calculated over the appropriate period of time. For emergency exposure situations where the dose is likely to be received in less than 1 year, the residual dose calculated and compared with the reference level should be the total dose received as a result of the emergency exposure situation. For accidents where the total dose is likely to be received over periods of more than 1 year, the residual dose calculated and compared with the reference level should be the sum of the external dose received over 1 year plus the committed effective dose received from intakes over the same 1-year period. With the exception of high-consequence, low-probability emergencies, the Commission recommends that if the residual dose exceeds the appropriate reference level, additional protective measures should be planned to result in residual doses less than the reference level. Where the implementation of protective measures is being considered during the actual response, the residual dose calculated for comparison with the reference level should include doses already received, those committed through the ingestion and inhalation of radionuclides, and those expected to be received in the future (see Section 8).

(28) The process of constrained optimisation should result in a residual dose which is below the appropriate reference level that is both radiologically and socially acceptable. This is because the process of optimisation involves wider issues than simply the radiation health risks associated with the dose. The process of optimisation must take account of the perceptions and aspirations of those who will continue to live and work in affected areas, and of those who may visit or purchase goods from these areas. Doses that are acceptable in the longer term will be influenced by the doses actually received. Therefore, it is generally the full 1-year residual dose

(dose received plus the prospective dose for the remainder of the year) that should be the object of an optimised protection strategy. The optimised outcome may also be influenced by other – non-radiological – measures taken to support those affected, e.g. compensation schemes, health monitoring, infrastructure, and economic support. Therefore, in the planning of a protection strategy, it is important to engage the potentially affected stakeholders and, to the extent possible, explore with them what overall outcome, including residual dose, would be acceptable. This process of constrained optimisation within emergency response planning is discussed in more detail in Section 7.

5.3. Averted dose

(29) The averted dose is the dose to the appropriate representative person (usually expressed as effective dose or equivalent dose) that is expected to be averted by the implementation of a protective measure or combination of protective measures. The concept of averted dose is an important component of the optimisation of emergency response planning, since it is one measure of the radiological benefit gained from implementing a protective option.

(30) Full response strategies are comprised of a set of individual protective measures (such as evacuation, milk restrictions, etc.). Where a number of different protective measures are relevant, optimisation of the overall protection strategy as a single process is likely to be complex. To assist emergency planners, the Commission has published guidance on setting intervention levels of averted dose for individual protective options (ICRP, 1991a,b, 2005). These are intended to assist in the optimisation of the component protective measures of an emergency response plan. It should be emphasised that the intervention level is understood as a level above which an action is justified and below which no action is needed. This concept is no longer valid. Thus, the above-mentioned intervention levels should be called ‘triggers’, expressing the effectiveness of a specific protective measure in terms of averted dose in view of its inclusion in an integrated protection strategy of multiple protective measures.

5.4. References

ICRP, 1991a. 1990 Recommendations of the International Commission on Radiological Protection. ICRP Publication 60. Ann. ICRP 21(1–3).

ICRP, 1991b. Principles for intervention for protection of the public in a radiological emergency. ICRP Publication 63. Ann. ICRP 22(4).

ICRP, 2005. Protecting people against radiation exposure in the event of a radiological attack. ICRP Publication 96. Ann. ICRP 35(1).

ICRP, 2006. Assessing dose of the representative person for the purpose of radiation protection of the public. ICRP Publication 101. Ann. ICRP 36(2).

ICRP, 2007. The 2007 Recommendations of the International Commission on Radiological Protection. ICRP Publication 103. Ann. ICRP 37(2–4).

6. APPLYING THE COMMISSION'S SYSTEM TO EMERGENCY EXPOSURE SITUATIONS

(31) The Commission's 2007 Recommendations (ICRP, 2007) re-state its principles of justification and optimisation, and the requirement to protect against severe deterministic injury, as applying to emergency exposure situations.

- The principle of justification: any decision that alters the radiation exposure situation should do more good than harm. This means that (in taking action) one should achieve sufficient individual or societal benefit to offset the detriment it causes.
- The principle of optimisation of protection: the likelihood of incurring exposures, the number of people exposed, and the magnitude of their individual doses should all be kept as low as reasonably achievable, taking into account economic and societal factors. This means that the level of protection should be the best possible under the prevailing circumstances, maximising the margin of benefit over harm. In order to avoid severely inequitable outcomes of this optimisation procedure, there should be restrictions on the doses or risks to individuals from a particular source in the context of emergency exposure situations; these restrictions are termed 'reference levels'.
- The requirement to protect against severe deterministic injury: situations in which the dose thresholds for severe deterministic injuries could be exceeded should always require action.

(32) In an emergency exposure situation, the source of exposure is not inherently under control, and its magnitude may vary over a large range. It may therefore be impractical to maintain exposures below a predetermined dose level, or, in order to achieve this, the emergency response may require the use of resources well in excess of the risk. As such, the Commission recommends that exposure management should rely on the concept of constrained optimisation rather than on the more rigid concept of dose limits.

(33) Exposures may occur in advance of the emergency being recognised (e.g. covert malicious actions or orphan source material incorporated inadvertently in commodities). Even following recognition of an emergency situation, there may be no practical protective measures that could reasonably be planned to reduce or avoid some exposures (e.g. the initial exposures to those in close proximity to a criticality accident). Furthermore, some emergency scenarios may be so unlikely to occur that detailed planning of protective actions would not constitute an acceptable use of resources. The Commission's advice on planning protection strategies for emergency exposure situations is only intended for application to those aspects of exposure and protective response for which it is reasonable to plan. It is the role of national authorities to set the regulatory framework for identifying the emergency scenarios for which it is appropriate to develop emergency response plans, and those aspects of the scenarios for which it is not reasonable to plan protective actions that optimise exposures below the appropriate reference level. The principles of justification and optimisation are discussed in more detail below.

6.1. Justification

(34) Protection strategies are comprised of a series of specific protective measures designed to address, as appropriate, all pathways to which affected populations may be exposed. This concept represents an evolution from the previous ICRP recommendations which suggested that the individual and independent justification and optimisation of individual protective measures was sufficient. The Commission now considers that more complete protection is offered by simultaneously considering all exposure pathways and all relevant protective measures when deciding, during planning, on the optimum course of action to be taken. In more concrete terms, this means that the overall 'benefit' and 'harm' of a suite of protective measures must be assessed when judging the justification of their application – it will be justified to implement a protection strategy when it results in more benefit than harm. In many cases, the summation of benefit and harm from a series of justified individual protective measures will also result in a net benefit. However, in some cases, particularly for large-scale accidents, the addition of many protective measures, each with a positive net benefit but each resulting in significant social disruption, could result in the collective benefit of the protection strategy being negative. Thus, while each individual protective measure must itself be justified, the full protection strategy must also be justified, resulting in more good than harm.

(35) As an emergency exposure situation evolves, prevailing circumstances may change beyond the range that was considered during emergency planning and preparations. As such, the protection strategy may also require change. In considering these changes, the 'new' protection strategy should be justified. The level of detail given to such 'real-time' justification will, of course, vary depending on the urgency of the situation at hand, but it would be expected that beyond the need for urgent protective measures, a careful assessment of the justification of proposed actions would take place, as well as their subsequent optimisation. The need to reconsider justification will be a judgemental decision based on the magnitude of any necessary changes to the original plan.

(36) The resources allocated to the process of justification will vary depending on a number of factors. Two of the most important factors are the nature of the likely health effects should an emergency exposure situation occur, and the extent to which the need for protective measures can be 'deferred' until the occurrence of the exposure situation (i.e. whether the planned response can largely rely on 'paper' planning and training, or whether specialised equipment, e.g. alarm systems, must be purchased and/or installed in advance). The Commission also recognises that the practicability of implementing a protective measure is relevant to determining whether or not an action should be included in the protection strategy.

6.2. Optimisation and the role of reference levels

(37) When optimising protection strategies, it is necessary to consider all aspects and protective measures to reduce residual dose, questioning whether '...the best has been done in the prevailing circumstances, and if all that is reasonable has been

done to reduce doses' (ICRP, 2007, Para. 217). This approach focuses efforts on optimising protection in order that individual exposures, from all pathways, resulting from the emergency exposure situation (i.e. residual doses) are judged to be acceptable in the context of the circumstances being planned for and the expected resources required/allocated for protection. This new approach implies the simultaneous optimisation of all protective measures that are included in the protection strategy, implemented if necessary in a stepwise fashion to address prevailing circumstances appropriately.

(38) While this new approach does represent a relative increase in operational complexity, it also provides a significant amount of increased flexibility in designing the 'best' protection to address an emergency exposure situation. This is partly because it enables the influence of one protective measure upon another to be taken into account, and partly because it provides for resources to be focused towards those measures that are expected to achieve the greatest net benefit overall, rather than implying the need to focus equal attention on each single protective measure. The sum of the benefits and harms from all individual, optimised protective measures may not, itself, be positive. Again, this is because the combination of effects of many individual protective measures, each involving large social disruptions, may be too socially disruptive overall. An optimised protection strategy may include protective measures which would not appear optimised if considered in isolation.

(39) The Commission has introduced the concept of constrained optimisation below reference levels in order to ensure that the response, as well as being optimised, avoids inequity of individual exposures. The reference levels define a level of dose above which it is generally unacceptable to plan to allow exposures to occur, and below which one should strive to be as low as reasonably achievable. The level of dose applies to that estimated for representative persons of identified population groups.

(40) In planning, it is necessary to optimise protection with respect to specific population groups and with respect to the overall response. This latter consideration is illustrated in Fig. 6.1. In this figure, the vertical lines represent the spread of residual doses to the representative persons of all population groups considered in the plan, whilst the blocked bars represent the mean of these residual doses. In this case, only

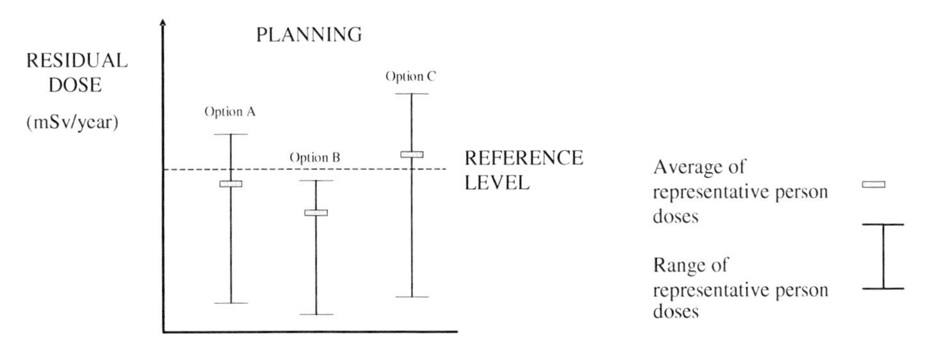

Fig. 6.1. The application of dose reference levels in planning protective actions for a number of population groups: Options A and C are unacceptable.

Option B is acceptable, as Options A and C result in doses to representative persons that exceed the reference level.

(41) In order to optimise an overall protection strategy during planning, it is necessary to identify the dominant exposure pathways, the time scales over which components of the dose will be received, and the potential effectiveness of individual protective measures. Knowledge of the dominant exposure pathways will guide decisions on the types of protective measures to consider and the allocation of resources; resources allocated to protective measures should be commensurate with the expected benefits, of which averted dose is an important component. Knowledge of the time scales over which exposures will be received informs decisions about the lead times available to organise protective options once an emergency exposure situation has been recognised. The effectiveness of single protective measures can be complex to evaluate, as it includes wider social and economic consequences as well as dose effectiveness.

(42) The optimisation of the overall protection strategy during planning is an iterative process involving stakeholders, in which the proposed protective measures are individually optimised and their contribution to the overall protection strategy is assessed and optimised. For planning, this optimisation needs to be robust as the detailed circumstances of the emergency cannot be known in advance. When an emergency has been recognised, the appropriate protection strategy should be implemented. Once the urgent measures have been implemented, a more detailed iterative optimisation can take into account the exact circumstances and the actual stakeholders. Thus, the process of constrained optimisation is iterative with respect to individual measures and the overall protection strategy, with respect to time and stakeholders. At each stage, comparison of the residual dose expected from the overall protection strategy should be compared with the planned residual dose to gauge

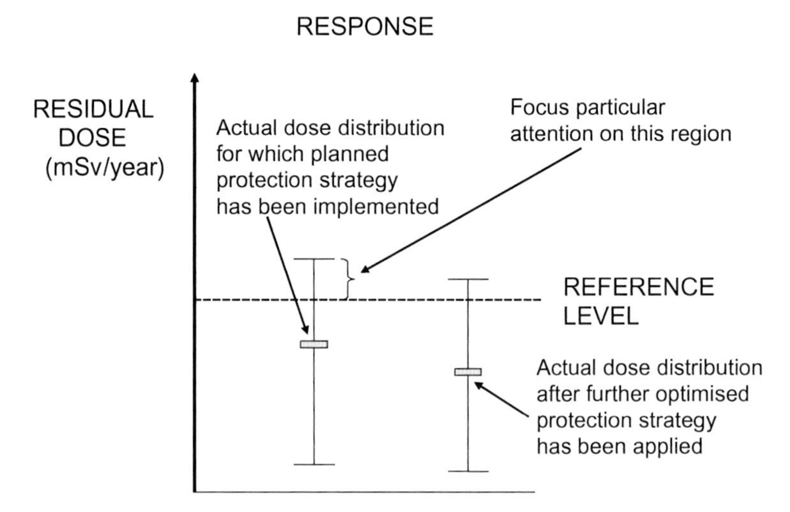

Fig. 6.2. Actual dose distribution after implementing the planned protection strategy (left) and ongoing optimisation (right).

the effectiveness of the implemented protection strategy, as well as the appropriate reference level(s) to ensure that the outcome is optimised.

(43) Fig. 6.2 illustrates the application of the reference level once an emergency situation has occurred. Regular review of the expected residual doses as both the emergency situation and the response develop, and consequent re-optimisation of the response may well result in a progressive lowering of the expected residual doses over time. Review of the expected residual doses may also demonstrate that the doses to some population groups will exceed the reference level. In this case, any re-optimisation of the protection strategy should focus on these population groups in order to explore whether it is feasible to reduce their doses. It should be noted, however, that a fully optimised response may result in a distribution of doses where some are above the reference level, as indicated by the second bar in Fig. 6.2.

6.3. Reference

ICRP, 2007. The 2007 Recommendations of the International Commission on Radiological Protection. ICRP Publication 103. Ann. ICRP 37(2–4).

7. ARRANGEMENTS FOR EMERGENCY EXPOSURE SITUATIONS

7.1. The planning process

7.1.1. Preparing response plans

(44) The importance of planning for emergency response cannot be over-emphasised. No emergency response can be effective without prior planning. This planning needs to involve identification of the range of different types of emergency situation for which a response may be required, engagement with stakeholders, selection of appropriate individual protective measures and development of the overall protection strategy, agreement of the areas of responsibility of different agencies and how they will interact and communicate, deployment of the necessary equipment for monitoring, supporting the implementation of protective measures, communicating with those at risk, training, and exercising of the plans.

(45) The Commission recommends that plans should be prepared for the types of emergency exposure situation identified in the risk assessment: nuclear accidents (occurring within the country and abroad); transport accidents; accidents involving sources from industry and in hospitals; malicious uses of radioactive materials; and other events. The level of detail within the plans will depend on the level of threat posed[1], and the degree to which the circumstances of the emergency can be determined in advance. However, even outline generic plans should indicate the responsibilities of the different parties involved, methods for communication and co-ordination between them and internationally during the response, and a framework for guiding decision making. More detailed plans should contain a description of the protection strategies for the identified scenarios, and provide triggers to facilitate prompt decision making.

(46) Detailed response plans are likely to place most emphasis on the initial response, as this is when there is least time for developing the response in real time and when uncertainties concerning the overall situation, current exposures, and the likely evolution of exposures will be greatest. However, any actions (or inactions) taken during this phase will impact on what can be, or needs to be, done at a later stage. In addition, the particular characteristics later in an emergency exposure situation, e.g. with regard to the need for widespread monitoring, may mean that unless the protection strategy is addressed adequately during planning, it may not be possible to respond effectively in the event. The optimum protection strategy for emergency exposure situations must address a wide range of issues over a relevant time period. For this reason, the Commission's reference level for emergency exposure situations refers to the residual dose received/committed over 1 year unless otherwise specified (Para. 27). Optimising the planned protection strategy to aim for the maximum residual doses to be below this level and as low as reasonably achievable

[1] The assessed level of threat will involve consideration of both the probability of occurrence of the event and the consequences, should it occur.

requires consideration of the response across all phases (or at least the early and intermediate phases for large events). An emergency protection strategy should therefore address the response during all time periods. For an urgent response, the planned response would be set out in detail with triggers to help guide decisions on implementation. For the later response actions, it is likely that an outline protection strategy would be indicated, rather than a specific response being planned, together with a framework for developing the specific response at the time of the emergency, taking account of the actual circumstances. Whilst the form of planning for the later time periods would be different, and in some cases may be done as part of planning for existing exposure situations, it is important that this planning is undertaken in such a way as to provide confidence that residual doses over a whole year will not exceed the reference level.

(47) Even after an emergency exposure situation has occurred, there will be a need to plan subsequent actions, particularly as time advances and the urgent need to act dissipates and finally disappears. As such, there will be a continual need to reflect relevant experience in selecting, justifying, optimising, implementing, adjusting, and terminating protective measures. The emergency plan will have identified a set of protective measures, and planned their implementation to an appropriate level of detail.

7.1.2. Protective measures to avoid severe deterministic injury

(48) A key concern in the event of an emergency exposure situation is to keep the exposure to individuals from all pathways below the thresholds for severe deterministic health effects (ICRP, 1991a,b). In the event of an emergency, it is possible that some individuals may be exposed to radiation doses that are so high that, without prompt medical treatment, they will cause severe, irreversible injury to their health. The Commission calls these 'severe deterministic injuries', to distinguish them from deterministic tissue reactions which may be reversible or may only have a minor impact on the individual's health. The Commission continues to advise that practicable protective measures should always be planned to protect those individuals who would be at risk of suffering severe deterministic injury in the event of an emergency exposure situation. The following paragraphs provide additional advice on a framework to achieve this.

(49) In developing this framework, the Commission recognises that there is a qualitative difference between planning protection for emergencies and protection for unintentional events (e.g. orphan sources) and malicious acts. Accidents occur when events disrupt planned exposure situations. It is therefore possible to design additional safety precautions into a planned activity that would mitigate the doses received in the event of an accident. This is clearly not possible in the case of malicious acts, as such acts are planned deliberately to circumvent any protective measures that might be in place. The Commission recommends that the framework for protection in the case of accidents comprises two steps: one prior to any accident, and one in the event of an accident. The recommended framework for protection in the case of malicious acts may contain a 'prior' step for specific locations or

activities judged to be at particular risk, but will generally focus on the response phase.

(50) The Commission recommends that postulated emergency situations should be examined to determine whether or not they may result in exposures that could cause severe deterministic injuries. If such exposures are considered possible, all protective options that could be implemented in advance to reduce these exposures, should the emergency exposure situation occur, should be considered. Such options will be dependent on the particular circumstances and may include:

- engineering (e.g. additional shielding, containment, filtration, interlocks, alarm systems, separation distances of stored fissile material);
- procedures (e.g. restrictions on those who may enter a particular area, requirement for using personal protective equipment); and
- training (e.g. recognition of alarms and response to them, suitable qualification and experience to operate plant and equipment).

The Commission recommends that all options are presumed justified, and therefore implemented, unless a specific case can be made to demonstrate the contrary. Reasons for an option being considered unjustified may include:

- disruption of normal activities to an unreasonable extent;
- the placing of an unreasonable economic burden on the operation;
- the introduction of a greater risk by their implementation than they are designed to protect against; and
- that another protective option associated with a smaller risk or effort exists which provides the same or better protection.

However, it is important that every option is considered explicitly so that maximum practicable protection is provided.

(51) A further step is the preparation of a protection strategy that provides specific protection for those at risk of severe deterministic injury in an emergency exposure situation. The protection of those at high risk should take priority over the protection of others, in terms of both resources and focus. Therefore, this part of the overall response should be separate from the protection of those at risk of lower exposures. Nothing in the emergency plan developed for protecting those at lower risk should compromise the protection of those who are at risk of receiving exposures that could result in severe deterministic injury. The response plan for those at risk of severe deterministic injury should not only include specific measures aimed at reducing exposures, but should include procedures to rapidly identify those most likely to be at high risk, so that they can receive detailed assessment and prompt medical attention. One way of achieving this might be for those in a particular high-risk area to muster separately to others, so that they can be accorded appropriate priority.

(52) In planning protection against malicious acts, it may not be possible to implement protection in advance. However, for specific 'high-risk' locations and activities, the Commission recommends that practicable options should be considered for implementation, in order to reduce the exposures that may result from

such acts. Response planning for malicious acts should develop procedures that enable the potential for exposures sufficient to cause severe deterministic injury to be assessed rapidly. Where this potential is judged to exist, the emergency plan should also provide procedures for identifying those who may have been exposed at such levels, guidance on exposure assessment and appropriate treatments, and practical plans to enable the individuals to receive treatment in the necessary time scales.

7.1.3. Engagement with stakeholders (who to involve)

(53) The Commission recognises that the nature and extent of stakeholder involvement may vary between countries, but suggests that engagement with stakeholders is an important component to the justification and optimisation of protection strategies in emergency exposure situations.

(54) During planning, it is essential that the plan is discussed, to the extent practicable, with relevant stakeholders, including other authorities, responders, the public, etc. Otherwise, it will be difficult to implement the plan effectively during the response. The overall protection strategy and its constituent individual protective measures should have been worked through with all those potentially exposed or affected, so that time and resources do not need to be expended during the emergency exposure situation itself in persuading people that this is the optimum response. Such engagement will assist the emergency plans by not being focused solely on the protection of those at greatest risk early in an emergency exposure situation.

(55) Stakeholders are not limited to those groups affected in the country where the emergency occurs; for large-scale emergencies, there may be international consequences. These may result from: international trade and concerns that produce/trade items may be contaminated; the perceived need for protective measures in other countries and therefore the need to harmonise the response across country borders; and the need for authorities to ensure the safety of their nationals in an affected country and to deal appropriately with people from an affected country crossing their borders. It is important that national authorities ensure effective international communication with authorities, particularly in countries that could be affected in the event of an emergency. There would be advantages in co-ordinating the response as much as possible.

(56) A further need for stakeholder engagement centres around the issue of waste containing radioactive material. In any emergency exposure situation involving anything more than the most limited contamination of the environment, it is likely that very large volumes of contaminated waste will be generated, e.g. Goiania (IAEA, 1988). Early in an emergency exposure situation, emphasis should be placed on containment of radioactive waste so as to control impacts on people and the environment. In the longer term, management and disposal of waste will pose significant problems both socially and practically, and may even require changes to legislation. Where agriculture is affected, the problem of large volumes is compounded by the fact that this waste may rapidly become a health hazard, and the production of some

food wastes (e.g. milk) is not easily terminated. Engagement with representatives of local communities, producers, and regulators in advance of an emergency can provide an opportunity for solutions to be developed in outline, and any legislative changes required to be identified in advance.

7.1.4. Representative persons (who to protect)

(57) In the event of an emergency exposure situation, it is likely that actual exposure rates will vary in space and time, and that the doses received by individuals will vary, both as a result of the variations in exposure rates and as a result of differences in their physiological characteristics and behaviours. In order to ensure that the optimum protection strategy is developed, it is important to consider the range of doses and other consequences for individuals that may occur, both in the absence of protective measures (projected doses) and following implementation of the protection strategy (residual doses).

(58) The Commission advises that this should be achieved by identifying a set of different population groups who, by their locations, characteristics, and behaviours, represent the full distribution of doses and risks. These population groups should be characterised by representative persons, as described in the Commission's advice on representative persons (ICRP, 2006). It would be expected that where children and other sensitive groups are likely to be present in an affected area, the consequences and protection strategy for these groups would be explicitly considered as deemed appropriate in the planning arrangements. In accordance with the Commission's advice on the representative person, it is important that the dose estimates reflect those likely to be received by the groups at greatest risk, e.g. pregnant women and children, but that they are not grossly pessimistic.

7.1.5. Setting reference levels

(59) The Commission has recommended that reference levels for emergency exposure situations should typically be set in the band of 20–100 mSv (acute or per year). This band applies in unusual, and often extreme, situations where actions taken to reduce exposures would be disruptive (ICRP, 2007, Para. 241). Reference levels and, occasionally for 'one-off' exposures below 50 mSv, constraints could also be set in this range in circumstances where benefits from the exposure situation are commensurately high. Action taken to reduce exposures in a radiological emergency is the main example of this type of situation. The Commission considers that a dose rising towards 100 mSv will almost always justify protective measures. In addition, situations in which the dose threshold for deterministic effects in relevant organs or tissues could be exceeded should always require action.

(60) While reference levels for emergency exposure situations may be fixed at values up to 100 mSv (ICRP, 2007, Table 5), they would only be set at the upper end of the band 20–100 mSv in unusual or extreme circumstances where actions taken to reduce exposures would be very disruptive. Setting reference levels for

emergency workers is discussed in Section 4. Reference levels below 20 mSv may be appropriate for the response to events involving projected exposures below 20 mSv.

(61) The selection of a reference level should fit the type of emergency exposure situation and the protection strategy to which it will be applied. For example, in a large-scale release of radioactive material, the protection strategy will be an evolving set of protective measures aimed at addressing the particular circumstances of populations affected in different ways and to different levels, at different times, and in different places. Part of the role of those authorities responsible for requiring emergency response planning should therefore be to determine the most appropriate reference level for the emergency exposure situation under consideration. This may be achieved by setting a single reference level for all types of emergency, or alternatively, by setting an appropriate reference level to apply to each type of emergency. There may also be situations where it is not possible to plan to keep all doses below the appropriate reference level, e.g. high-consequence emergencies in which extremely high acute doses are received within minutes or hours. For these events, the Commission advises that measures should be taken to reduce the probability of their occurrence, and response plans should be developed that can mitigate the health consequences where practicable. Emergency planners should prepare their protection strategy in accordance with the established reference level.

(62) The preselected reference level against which the optimisation of protection is assessed should be expressed in mSv (acute or per year). The optimisation process may also need to take into account whether or not the optimum protection is being afforded to individuals under various emergency circumstances. The residual dose to be compared with the preselected reference level is that assessed and/or estimated for the exposed populations for the year following the accident. In an emergency exposure situation involving no long-term environmental contamination (e.g. criticality accident), the preselected reference level against which the effectiveness of the protection strategy is assessed is the total dose received via all pathways over whatever time period the exposures may occur.

(63) The type of reference level that is selected should thus be tailored to meet the type of emergency exposure situation under consideration. Depending on the time of year of a release to the environment, as well on the nuclide composition of a contamination, there are major differences in the contributions of the pathways considered (ingestion, inhalation, cloud shine, ground shine) to the projected dose. These differences have to be considered in an appropriate fashion in defining reference levels and developing protection strategies. Regulatory authorities and operators will assess reasonably foreseeable risks, and authorities will preselect appropriate reference levels for the various emergency scenarios that they judge to be relevant.

(64) The Commission's band of reference levels is expressed in terms of effective dose. There are situations for which effective dose is not an appropriate quantity in which to express reference levels. Such situations include: where the type or scale of the emergency may result in doses in excess of 100 mSv effective dose (where the

assumption of linearity in its derivation may no longer hold); where parts of the response need to focus on individuals at risk of incurring severe deterministic injury; and where the resulting exposures are very strongly dominated by irradiation of a single organ for which very specific protective measures are optimum (e.g. releases dominated by radioiodine). For these situations, the Commission advises that consideration should be given to specifying (or providing supplementary) reference levels in terms of equivalent or absorbed dose.

7.1.6. Role of intervention levels

(65) In its previous advice (ICRP, 1991a,b, 2005), the Commission recommended the use of intervention levels of averted dose to assist decisions on whether/when to include certain protective measures in an overall protection strategy. This term was used in the context of a process-based approach system of protection, making a distinction between practices and intervention. In this latter case, the concept of intervention level was understood as requiring an action above the intervention level and no action below; this is not consistent with the 2007 Recommendations (ICRP, 2007). Thus, the Commission believes that it is more appropriate to avoid the use of the term 'intervention level'. However, the corresponding quantified values can be used as triggers expressing the effectiveness of individual protective measures as an input of the process of optimisation of the protection strategy (ICRP, 2005).

(66) In the context of planning and implementing a protection strategy, the Commission still considers that the levels of averted dose are useful tools to rate the individual protective measures within an optimised protection strategy. However, in an actual emergency exposure situation, the parameters relevant to the situation may be different than the levels of averted dose for individual protective measures, and/or the judgements used in establishing these levels in isolation may not be fully representative. When several actions are combined within an overall protection strategy, the balance of harms and benefits contributed by individual measures is likely to be different than when they are considered in isolation. It is therefore not appropriate to treat the averted dose levels previously defined by the Commission as 'absolute' criteria that prescribe when each protective measure should be included in a plan. In particular, in circumstances for which it appears that no single protective measure on its own is sufficient to reduce residual doses to below the reference level and as low as reasonably achievable, it may be necessary to combine several protective measures, one or more of which would appear unjustified by simple comparison with its previously defined intervention level, to achieve this outcome. In this case, the level of averted dose would act as a prompt to consider the introduction of that measure more carefully, to determine whether or not an alternative measure or manner of introduction might increase the expected benefits or decrease the expected harms. Triggers, particularly their use for the initiation of urgent protective actions, are further discussed in Section 7.2.5.

7.2. Components of a protection strategy

(67) A protection strategy will contain a wide range of information and guidance, including information such as contact details, duties and responsibilities of the different organisations involved, reference to legislation, amounts of equipment/resources required etc.; this is beyond the scope of this document. A discussion of these practical and technical issues can be found in publications from other organisations (NEA, 2000; IAEA, 2002, 2003). In this document, only those aspects relevant to the application of the Commission's advice are discussed.

7.2.1. Strategies and individual protective measures

(68) There are different protective measures which could be applied in a radiation emergency. Urgent protective measures are those that must be taken promptly (normally within hours) in order to be effective, and for which the effectiveness will be markedly reduced if there is a delay. The most commonly considered urgent protective measures in a nuclear or radiological emergency are evacuation, decontamination of individuals, sheltering, respiratory protection, iodine thyroid blocking, and restriction of the consumption of foodstuffs that have the potential to give significant exposures to people (e.g. green vegetables grown in the open and milk from animals grazing outdoors). Longer term protective measures (and food restrictions to protect against longer term exposures) include measures such as permanent relocation, agricultural protective measures, and some decontamination measures. The Commission has previously published detailed guidance on most of these protective measures (ICRP, 1991a,b); further discussion of individual protective measures in this document is therefore restricted to new aspects of the Commission's advice.

(69) During an emergency exposure situation, other measures are also likely to be considered. These include public warning, information, advice and basic counselling, dealing with their own national citizens in another affected country, comprehensive psychological counselling, medical management, and long-term follow up. More details are provided in Annex B.

7.2.2. Temporal and geographical issues

(70) The characteristics of potential exposures and therefore the requirements for the protective response will vary both spatially and in time. In order to be manageable, a protection strategy will sub-divide the area at risk into appropriate sub-areas, based on a number of factors such as: distance from the initiating source; demographic, economic and land use factors; and response phase (early, intermediate, late). This approach enables the broad issues for each sub-area to be treated appropriately within the plan. However, in reality, there will be few, if any, sharp boundaries to delineate the implementation of protective measures.

(71) Aspects that need to be considered when optimising an overall protection strategy are: the impact of actions (or inactions) taken at one point in time on subsequent protective requirements; and the potential need to manage different areas

simultaneously in different fashions (e.g. one area with heavy contamination may require urgent protective actions, whilst another area with much lower contamination may require management involving more stakeholder engagement). These types of issues are discussed below.

(a) Influence of actions on subsequent actions

(72) The management of wastes arising from decontamination, food restrictions, and other protective measures (e.g. domestic and commercial refuse left outside in an area that was evacuated) is an example of the need for response planning to consider the wider temporal consequences of actions. A decision to prevent the consumption of fresh milk is straightforward to make and implement, but the consequence is a rapid build up of large volumes of organic liquid waste that are difficult to dispose of safely, from a biological perspective, regardless of the radiation hazard posed. An optimised overall protection strategy should include the identification and prior agreement of appropriate disposal routes and temporary storage sites.

(73) The termination of protective measures is another area where the interaction of urgent protective measures and later protective measures is particularly obvious. Withdrawing all urgent protective measures and then, some time later, initiating new protective measures such as decontamination may, purely from consideration of future doses and dose rates, seem the optimum course of action. It may not be optimum from a practical and 'cost' viewpoint. For example, extending evacuation whilst decontamination is carried out may not actually increase the monetary costs of the combined protective measures substantially, as decontamination may be carried out more efficiently in the absence of people living in the area.

(74) The Commission therefore advises that effective emergency response planning should consider the impact of protective measures taken at one time on decision options available at later times. It may be that one way of ensuring that these interactions are adequately addressed is for the planned response to identify the need for a team to be set up early in an emergency exposure situation, whose responsibility is primarily to consider what might be required later, and how early decisions might impact on this.

(b) Dynamic nature of response

(75) The spatial variation of future exposures inevitably results in some areas potentially having much higher exposures than others. In the case of release of radionuclides to the environment, it is likely that areas close to the source will experience higher levels of contamination than areas further away from the source. For large releases, particularly, it is likely that appropriate reassurance monitoring will enable areas with low contamination to move to a less urgent form of response management, whilst the more highly contaminated areas will remain subject to the protective measures and management approach planned for an urgent response. One consequence of this is that, depending on the plan and the overall emergency response management approach of the country, different authorities may be responsible for

management of the different areas. Whilst simultaneous management of the response in different areas may not be problematic in principle, people living and working near the boundaries of such areas, or even living in one such area and working in another, may require particular care and involvement in the decision process. It will thus be important to foresee such situations in planning in order to avoid serious operational and social problems.

7.2.3. Developing a protection strategy

(76) In order to develop a protection strategy, it is necessary to evaluate the projected doses for the situation being considered. Only for the very largest emergency scenarios are projected doses received over time scales of fractions of a second up to 1 or 2 days likely to exceed 100 mSv (see Annex A). However, there is a wider range of postulated emergency scenarios which may result in projected doses in excess of 100 mSv effective dose from intakes over the first year, together with external doses received over that period.

(77) Examples for the relative contribution of different pathways (cloud shine, inhalation, ground shine, ingestion) to the projected dose are given in Fig. 7.1 for an atmospheric release of five radionuclides (^{60}Co, ^{90}Sr, ^{131}I, ^{137}Cs, and ^{239}Pu) and for two release scenarios for nuclear power plants (for details, see Annex A).

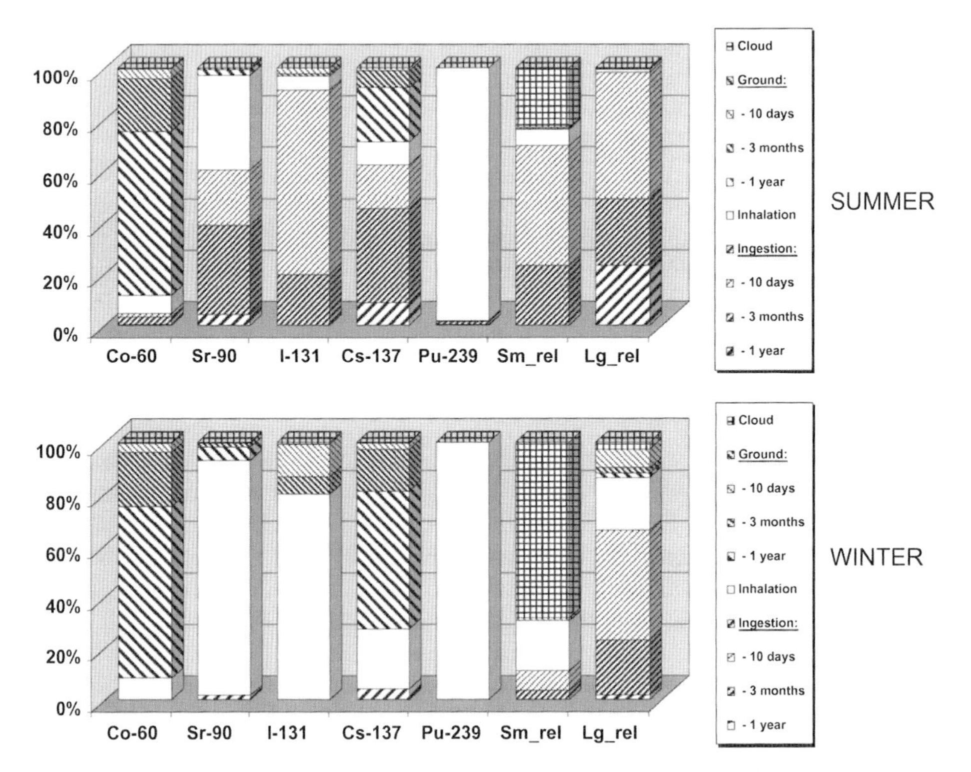

Fig. 7.1. Examples for the relative contribution of different pathways to the total dose.

(78) By contrast, the spread of contamination experienced in London following the poisoning of Mr. Litvenenko in November 2006 resulted in actual or possible acute intakes of an alpha-emitting radionuclide, via inhalation of resuspended particles or ingestion, but no risk from external irradiation nor contamination of foodstuffs (except secondary contamination spread from contamination on hands or contaminated utensils).

(79) The purpose of estimating projected doses and their likely spatial and temporal distributions is three fold: (1) to identify the scale of health consequences that might occur if no protective measures are taken and, therefore, to determine the broad scale of resources it is appropriate to assign to a protection strategy; (2) to identify the broad geographical and temporal distribution of the various likely response phases; and (3) where, in terms of protection, resources are likely to be spent most effectively. Identifying these broad trends helps to highlight the evolving response management issues that need to be addressed, and to provide initial guidance on the sorts of protective measures that might be appropriate. In respect of the third aim, it is likely that protective measures aimed at reducing the dose from the exposure pathway(s) that would otherwise dominate the projected dose will have the potential to avert the greatest dose. It is therefore reasonable to allocate resources to the detailed evaluation of such protective measures. In the context of particulate releases from nuclear reactors, it is likely that ingestion doses will dominate the projected doses over the first year; as a consequence, protective measures on the food chain are likely to have the potential to avert the greatest dose.

7.2.4. Detailed optimisation

(80) The timing and manner of implementation of individual protective measures and how they are combined into a protection strategy will influence the overall net benefit achieved by the protection strategy. It is therefore important to optimise the broad protection strategy. In this context, the Commission's previous advice on individual protective measures and their optimisation is relevant (ICRP, 1991a,b, 2005).

(81) In principle, the process of applying constrained optimisation to the planning of strategies of protective measures is the same as that for individual protective measures, i.e. all the consequences, harmful and beneficial, expected to result from the imposition of different strategies are evaluated and balanced, and the one with the greatest net benefit, which also results in a residual dose below the reference level, is selected. However, in practice, the problem with this process is that there are so many combinations of strategies that could be considered that the process could quickly become too complex. It is therefore advisable to adopt a more pragmatic approach, in which individual protective measures are optimised separately, and issues associated with their application in combination are identified and explored, as discussed below. A protection strategy composed of individually optimised protective measures will not necessarily be optimised itself, whilst an optimised protection

strategy may contain actions implemented in a way that, taken in isolation, would not be optimum.

(82) Some combinations of protective measures can be considered to be largely independent of each other, e.g. commercial food restrictions and the evacuation of populations in close proximity to a radiation source. The actions and resources required to implement these protective measures are very different (buses and evacuation centres for evacuation, food monitoring equipment and facilities for disposing of or processing foods in the case of food restrictions), and the impacts (other than doses averted) resulting from implementing these protective measures are likely to be sustained by different population groups (those evacuated, voluntary groups providing food and bedding, bus drivers for evacuation, farmers, food manufacturers, agencies responsible for monitoring food production and disposing of wastes in the case of food restrictions). These types of protective measures can readily be optimised separately and the relevant levels of averted dose used as a direct guide.

(83) With other combinations, the protective measures are more closely linked, with actions required to implement one being relevant for implementing the other. In this case, there is the potential for significant interaction between the harms (including resources required) and benefits of the options, and so the process of detailed optimisation is less straightforward. In this case, single measure levels of averted dose need to be used more flexibly, taking into account both the enhanced benefits of combining actions and, as above, the need to develop a plan that reflects the characteristics of the surrounding area, such as geographic and demographic areas.

(84) The resources required to implement protective measures are not the only factors that may interact within an overall protection strategy. Other such factors include individual and social disruption, anxiety and reassurance, and indirect economic consequences. It is important to review the proposed overall protection strategy with representatives of all potential stakeholder groups to ensure that the plan is optimised and feasible with respect to these factors, as well as with respect to dose and the resources required. This wider review of the protection strategy may indicate a role for additional actions which, in isolation, do not appear optimum (or even justified). Alternatively, it may indicate that the optimum protection strategy should modify or omit other actions, despite the fact that they appear to be justified and optimised when dose and direct resource requirements alone are considered.

(85) The level of detail of optimisation of the response plan should be commensurate with the needs of the situation. This is a matter for national authorities to decide. In general, events with the potential to result in widespread, high levels of contamination and events that are relatively likely to occur require more detailed planning than very-low-probability events or events that are expected to have limited consequences.

(86) At stages during the detailed optimisation of the protection strategy, the expected residual doses should be compared with the reference level to ensure that the outcome of optimisation remains below this level. Development of the plan is

therefore an iterative process, with the degree of iteration depending on the level of detailed optimisation considered appropriate for the significance of the emergency exposure situation and the need to provide for flexibility of response on the day.

7.2.5. Triggers

(87) Once the protection strategy has been optimised in planning, measurable triggers for initiating its different parts should be developed. Since most protective measures taken early in an emergency exposure situation need to be taken promptly, any delay in decision making would be counterproductive. Therefore, protection strategies should include triggers (IAEA, 2002) that can be used immediately and directly to initiate appropriate protective measures. Once an emergency is occurring, the types of information likely to be available to the decision maker will change with time, e.g. from assessments of plant conditions and limited dose rate measurements, to widespread and increasingly detailed information based on a substantive monitoring programme.

(88) Triggers may be expressed in terms of any observable circumstances or directly measurable quantities, e.g. plant conditions, dose rates, and wind direction. Triggers may be related to dose considerations, but are more likely to be quantities or qualities, such as information that a filter or a pump has failed, which indicate that a situation has occurred for which the plan (or a group of actions within the plan) was developed. Similarly, triggers may be identified which indicate that the event is outside the range of scenarios considered when the plan was developed, thus warning decision makers that the scale of protective measures may need to be escalated from those set out in the plan (in particular, the areas over which urgent protective measures are introduced may need to be enlarged significantly). Once the occurrence of a trigger has been identified, decision makers can advise that the appropriate part of the protection strategy should be implemented immediately, without further delay or discussion.

(89) In order to ensure widespread compliance with protective measures implemented on the basis of triggers, both with respect to those implementing the actions as well as those affected by them, it is important that relevant stakeholders (or their representatives) are involved at the preparedness stage in determining what the appropriate triggers should be. Unless this is achieved, the implementation of prompt actions on the day may be delayed whilst different groups demand further information to assure themselves that this is the best course of action.

(90) In some emergency exposure situations, it may become apparent that protective measures are required that were not envisaged within the plan, or for which the actions implemented were not sufficiently protective. In this case, decision makers should first implement all those urgent actions indicated by the triggers, but may then take additional actions not indicated by the planned triggers. In other words, the triggers should be used to facilitate prompt decision making, but not to prevent necessary flexibility to optimise on the basis of the exact circumstances of the emergency. This is discussed further in Section 8.

(91) Established criteria or indicators may also be helpful for deciding and delineating the extent of later protective measures. For example, once the radionuclide composition of environmental contamination is understood, a dose rate criterion could be applied to delineate where temporary relocation would be advised. Whilst specification of the triggers themselves in the emergency plan may not be appropriate, it may be helpful to include an agreed framework for developing triggers in 'real time'. The inclusion of such a framework is likely to result in wider acceptance of the 'real-time' triggers when they are developed.

7.3. References

IAEA, 1988. The Radiological Accident in Goiania. International Atomic Energy Agency, Vienna.

IAEA, 2002. Preparedness and Response for a Nuclear or Radiological EmergencyIAEA Safety Standards Series No. GS-R-2, Safety Requirements. International Atomic Energy Agency, Vienna.

IAEA, 2003. Method for Developing Arrangements for Response to a Nuclear or Radiological EmergencyEPR-Method. International Atomic Energy Agency, Vienna.

ICRP, 1991a. 1990 Recommendations of the International Commission on Radiological Protection. ICRP Publication 60. Ann. ICRP 21(1–3).

ICRP, 1991b. Principles for intervention for protection of the public in a radiological emergency. ICRP Publication 63. Ann ICRP 22(4).

ICRP, 2005. Protecting people against radiation exposure in the event of a radiological attack. ICRP Publication 96. Ann. ICRP 35(1).

ICRP, 2006. Assessing dose of the representative person for the purpose of radiation protection of the public. ICRP Publication 101. Ann. ICRP 36(2).

ICRP, 2007. The 2007 Recommendations of the International Commission on Radiological Protection. ICRP Publication 103. Ann. ICRP 37(2–4).

NEA, 2000. Monitoring and Data Management Strategies for Nuclear Emergencies. Nuclear Energy Agency, Paris.

8. IMPLEMENTING PROTECTION STRATEGIES

(92) In the context of the ICRP system of radiological protection, there is at least one fundamental difference between prospectively planning to address the consequences of a radiological emergency exposure situation, and managing consequences that are in the process of occurring or that have already occurred. In the context of planning, optimisation is performed using the appropriate reference level as the upper bound to optimisation, eliminating any protective solutions that result in individual residual doses exceeding the reference level. Protection for all those exposed is optimised, and residual exposures resulting from the application of the protection strategy are below the reference level. In the inherently unpredictable nature of emergency exposure situations, some actual exposures may exceed the preselected reference level. As such, in the context of managing the consequences of an emergency that is in the process of occurring or that has already occurred, the predefined reference level is used as a benchmark against which to judge the results of implementing a planned protection strategy, and for guiding the development and implementation of further protective measures if necessary (see Fig. 6.1).

(93) When reviewing the effectiveness of an ongoing response or making decisions on the implementation or variation of protective measures, it is important to compare the sum of all components of the residual dose with the reference level, i.e. the sum of doses actually received and those expected to be received in 1 year (or during the time frame relevant to the emergency exposure situation) with those expected in the planning stage.

8.1. Tuning protection strategies to actual conditions

(94) It is likely that if an emergency exposure situation occurs, it will not exactly match the assumptions that were made in planning. Divergence from initial assumptions will most likely increase as an emergency exposure situation progresses in time. However, in most cases, emergency response planning will broadly fit a large range of possible situations, such that the early and urgent implementation of a planned protection strategy should come close to providing the optimum protection, with divergence most likely being on the conservative side. However, there will still be some need to operationally adjust planned protection strategies, justifying any new actions or significant changes to the plans. The need to consider making such modifications will increase as the emergency exposure situation progresses, and the magnitude of changes from the plans will depend on the nature of the emergency exposure situation that occurs (i.e. large and complex, or small and straightforward). It is important, however, that minor modifications to the planned actions are avoided, particularly in the early period as the situation is developing and where uncertainties are greatest, as this is likely to, at best, introduce confusion for very little expected benefit, and at worst, actually result in reduced protection.

(95) Once an emergency exposure situation has occurred, it is likely that many stakeholders will be very interested in providing input to discussions regarding protective measures. Should the emergency exposure situation have an early period requiring urgent protective measures, the 'reflex' use of preplanned protection strategies will be necessary with no or very little stakeholder involvement beyond the emergency response authorities and those responsible for the site, facility, or source that is causing the emergency exposure situation. As the emergency exposure situation progresses, however, stakeholders will become increasingly interested and available to participate in discussions leading to protection decisions, such that part of emergency response planning should be the development and implementation of processes and procedures to inform and involve stakeholders.

(96) As an emergency exposure situation progresses, emergency planners may wish to approach optimisation in a stepwise manner, particularly for emergency exposure situations affecting large areas and/or causing long-lasting consequences. Here, periodic re-assessments may indicate that reference levels could be altered, generally in a downward fashion, to best assist the optimisation process as prevailing exposure circumstances evolve.

8.1.1. Tuning protection strategies early in an emergency exposure situation

(97) The early period of an emergency exposure situation can be characterised as following preplanned actions as best as possible to manage any emergency consequences. The focus of protection strategy decisions will be on adapting pre-made plans to best fit the actual situation.

(98) In the initial uncertain period of an emergency exposure situation, the radiological protection objectives of a protection strategy should be to avoid severe deterministic effects and to keep the risk of stochastic effects as low as reasonably achievable. To accomplish this, there may be a need to act very quickly and without much 'concrete' knowledge of exposures. Such 'reflex' protective measures will, of necessity, follow preplanned scenarios using preplanned procedures and processes. The plan should be developed such that revision of the most urgent protective options would not be required under almost any circumstances. However, re-assessment of the planned response in the light of actual conditions may be necessary after these most urgent measures have been implemented, so that actions best apply to the situation at hand. Re-assessment of the planned urgent response will involve as much specific information concerning the nature of the emergency exposure situation and its possible effects as possible, and any significant deviations from planning assumptions (i.e. extreme weather conditions, unexpected geographical location of the release site, temporary changes in population density due to unexpected circumstances such as large sporting or political events, etc.). In general, any modifications that are made to the planned response will be to extend protective measures in time and space.

8.1.2. Tuning protection strategies later in an emergency exposure situation

(99) As an emergency exposure situation progresses, and understanding of the exact circumstances increases, decisions will increasingly be based on actual circumstances rather than preplanned responses. As understanding increases and the need to act becomes less urgent, there will also be an increased need to plan future protection strategies in greater detail than included in the plan, and thus to involve relevant stakeholders in decision-framing and decision-making processes when judging the justification of protection strategies, and when optimising their application. For this planning of future action, a predetermined reference level will be a useful tool to deal appropriately with the situation at hand. The end-point of optimisation processes will be at least partially characterised by residual dose, which will have to be agreed by government (e.g. at the local, regional, and national level, and among relevant ministries) and relevant stakeholders (e.g. affected populations, affected businesses, etc.), and can be compared against the predetermined reference level when judging the appropriateness of the protection strategy.

(100) If, in application, protection options do not achieve their planned residual dose objectives, or result in exposures exceeding reference levels fixed at the planning stage, a timely re-assessment of the situation is warranted to understand why plans and results differ so significantly. New protection options could then, if appropriate, be selected, justified, optimised, and applied.

(101) Once protective measures start to be implemented, it is important to review their actual performance against the outcomes expected when the plan was developed. This feedback of actual outcomes and experience should be used to inform the further implementation of protective options and decisions on modifications to the later phases of the emergency plan.

(102) As an emergency exposure situation progresses and the need for urgent decisions dissipates, the decision-making process will inevitably shift away from giving direction towards furnishing an appropriate dialogue process with affected stakeholders so that the optimum protection strategy can be identified and implemented, and such that feedback of experience can help to improve the implementation of such protective measures. To incorporate stakeholder input appropriately into decisional processes, it is essential that structures, processes, and procedures, and perhaps legislation and regulation, are appropriately tuned to allow and encourage such participation.

(103) The active participation of stakeholders will, in general, bring relevant local knowledge, experience, and values to decision-making processes such that the resulting detailed protection strategies are more likely to be well focused, understood, and supported. However, the effective involvement of stakeholders will require appropriate training of the relevant staff from government bodies dealing with the emergency exposure situation in the social and interpersonal aspects of stakeholder involvement, bringing their technical knowledge to the service of the broader decisional process. In the long term, however, as the emergency exposure situation transitions

to an existing exposure situation, ongoing stakeholder involvement should become self-standing and independent.

8.2. Termination of protective measures

(104) The decision to terminate protective measures will need to appropriately reflect the prevailing circumstances of the emergency exposure situation being addressed. Many different aspects must be taken into account when reaching such decisions. Terminating a protective measure may result in the affected population receiving a 'step change' increase in dose rate, e.g. if access restriction to a contaminated area was terminated. However, planned protection strategies should consider the impact of termination of protective measures on the residual dose. In general, optimisation of the overall protection strategy during planning and reactive 're-optimisation' of the protection strategy throughout the emergency response will mean that the impact of termination of protective measures on the residual dose will have been factored into decisions. Therefore, in general, it would not be expected that the residual dose will change significantly as a result of the termination of protective measures. However, there may be unforeseen circumstances that require termination of protective measures at a time or in a manner that was not planned for. In this case, the resulting exposures following termination of the measures may be higher than initially planned, and may even exceed the reference level. As with decisions on the initiation of protective measures, whilst the requirement for optimisation of the overall protection strategy remains valid, the focus for decision makers should be on implementing further practicable measures to reduce the exposures of those population groups receiving doses in excess of the reference level. These points are illustrated in Fig. 8.1.

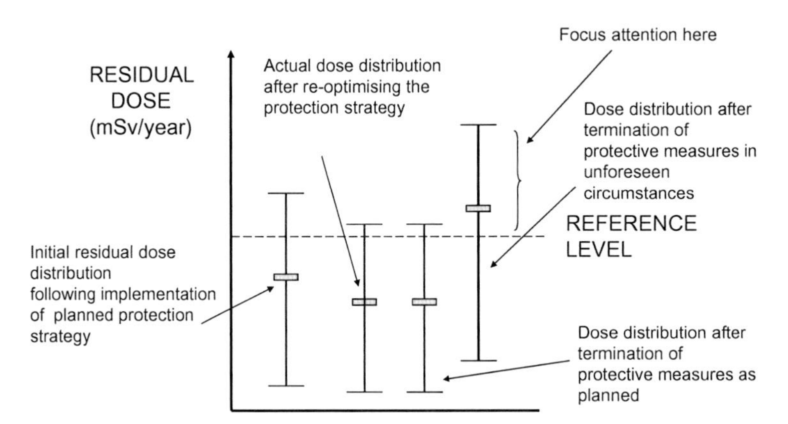

Fig. 8.1. Actual dose distribution after applying the planned protection strategy and optimisation (left) followed by the termination of a protective measure (right).

(105) It will be important to assess the potential benefits and harms caused by termination of a protective measure, and how this will affect the overall objectives of the protection strategy. A summary of issues which require consideration and evaluation before such decisions are made is given in Annex C.

(106) It is important to involve, wherever possible, relevant stakeholders in discussions regarding termination of protective measures. While it will be difficult, if not impossible, to discuss decisions with populations sheltered at home, it will be essential to discuss decisions to return to evacuated areas with those who have been evacuated, and the termination of protective measures implemented at a later stage.

(107) The types of information that will be needed for these decisions will, of course, vary between situations. However, in general, it will be important to have sufficient technical data on hand to judge the effects of termination. For example, the decision to allow evacuated populations to return to their homes and offices, i.e. the termination of temporary relocation, should only be taken when the exposures that would result from returning can be assessed appropriately, and this will most likely require an adequate understanding of the contamination conditions. For urgent protective measures that have short-natured effects, it will be essential to consider how their termination will affect the overall protection strategy and the optimisation process.

(108) Decisions regarding the termination of later protective measures will usually be based on the achievement of an optimum level of protection. These decisions will, in general, not be urgent in nature, and will be based on radiological protection input, and social and political judgement. In these cases, the involvement of relevant stakeholders is essential, and processes and procedures should be established to ensure that such involvement can take place efficiently.

(109) One important aspect of the involvement of stakeholders is that, when agreeing on the detailed implementation of any protective measure later in an emergency exposure situation, directly measurable outcomes should also be agreed (e.g. residual contamination levels, dose rates). This will help to ensure that when the action has been completed, it can readily be demonstrated that it has achieved the intended level of protection.

8.3. Permanent relocation

(110) For some large emergency situations involving the release of high levels of long-lived contamination over large areas, part of the new reality following the situation may be that some areas will be so contaminated as to be incapable of sustaining social, economic, and political inhabitation as previously. In these areas, governments may prohibit human habitation and other land uses. This would mean that any populations evacuated from these areas would not be allowed to return, and that further resettlement or use of these areas would not be allowed.

(111) Clearly, it is not easy for a government and its people to make a decision to permanently (or for the long-foreseeable future) remove people from an area and to forbid its use. As such, the social, economic, political, and radiological aspects of such a choice will need to be discussed in a broad and transparent fashion before

a decision is reached (IAEA, 1996, NEA, 2006). Generally, radiological aspects (e.g. contamination level, dose rate, etc.) would be used as part of the criteria to delineate the boundary of such areas, although existing geographic or jurisdictional boundaries may also be considered for social reasons.

(112) In corollary to a decision to define an area of permanent relocation is the fact that beyond such an area, people will be allowed to live. However, lingering contamination may well exist and may require the long-term management of population exposures. Transition from an emergency exposure situation to what is termed an 'existing exposure situation' is described in Section 9.

8.4. References

IAEA, 1996. One Decade After Chernobyl: Summing up the Consequences of the Accident. IAEA/WHO/EC International Conference. International Atomic Energy Agency, Vienna.

NEA, 2006. Stakeholders and Radiological Protection: Lessons from Chernobyl 20 Years After, OECD, Paris.

9. TRANSITION TO REHABILITATION

(113) At some point in time, the emergency situation will end. However, for a major accident resulting in the release of radioactive materials at a nuclear site, or a serious malicious contamination incident, some significant residual contamination of the environment may persist for a long period of time and continue to affect the population. The Commission recommends that the management of long-term exposures resulting from emergencies should be treated as an existing exposure situation.

(114) Existing exposure situations resulting from emergencies are characterised by the need for a population to continue living in an area with known or assessable levels of exposure. Typically, such situations have social, political, economic, and environmental aspects that are seen as sustainable by the affected populations and governments, and as being their new reality. Whilst there will not be a sharp transition, the early and intermediate characteristics of an emergency exposure situation will be distinct from those in the subsequent existing exposure situation. This is illustrated schematically in Fig. 9.1.

(115) The change from an emergency exposure situation to an existing exposure situation will be based on a decision by the authority responsible for the overall response. The decision may include considerations of the fact that different geographic areas may undergo this transition at different times. The transition may require a transfer of responsibilities to different authorities. This transfer should be undertaken in a co-ordinated and fully-transparent manner, and be agreed and understood by all parties involved. The Commission recommends that planning for the transition from an emergency exposure situation to an existing exposure situation should be undertaken as part of the overall emergency preparedness, and should involve all relevant stakeholders.

(116) There are no predetermined temporal or geographical boundaries that delineate the transition from an emergency exposure situation to an existing exposure situation. In general, a reference level of the magnitude used in emergency exposure situations will not be acceptable as a long-term benchmark, as these exposure levels

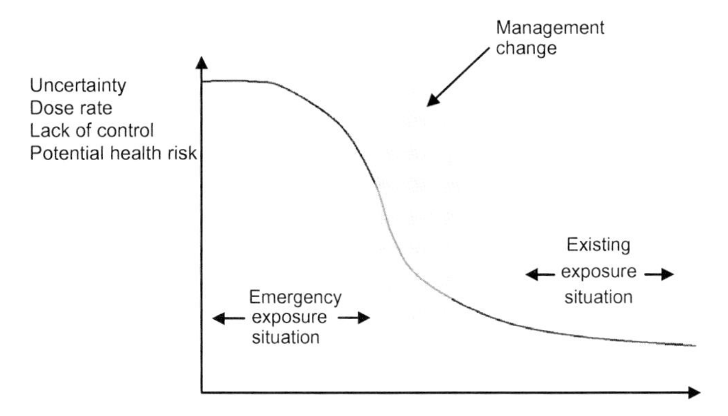

Fig. 9.1. Evolution of an emergency exposure situation with time and the transition to the existing exposure situation.

are generally unsustainable from social and political standpoints. As such, governments and/or regulatory authorities will, at some point, identify a new reference level for managing the existing exposure situation, typically at the lower end of the range recommended by the Commission of 1–20 mSv/year.

(117) If the level of contamination is such that sustainable social, economic, and environmental conditions cannot be achieved through the protection strategies implemented, the authorities may choose not to allow populations to live in some affected areas. The decision to permanently relocate populations will be based on radiological, social, and economic considerations with due recognition of the gravity and irreversibility of such a difficult decision.

(118) The management of existing exposure situations following an emergency will generally rely on the continuation and evolution of protection strategies implemented during the emergency, and the increased reliance on individual/self-help protection actions supported by an adequate infrastructure established by authorities. Generally, further significant reduction of exposures will not be achieved rapidly, but step-by-step optimisation will progressively bring exposures close to or similar to those associated with normal situations.

(119) More detailed guidance for the management of the long-term rehabilitation of contaminated areas following a nuclear accident or a radiation emergency is discussed in the complementary forthcoming ICRP publication.

ANNEX A. ASSESSMENT OF THE CONTRIBUTION OF DIFFERENT EXPOSURE PATHWAYS TO THE PROJECTED DOSE

(A1) In the case of a severe accident with a radioactive source or in a nuclear reactor resulting in an accidental atmospheric release, it is likely that projected doses will be characterised by an initial, relatively high dose rate, inhalation component from inhalation of short-lived beta/gamma emitters during dispersion of the plume. For a reactor accident, this is likely to be followed by a time period lasting days or weeks when I-131 dominates the exposures, through external irradiation from contamination deposited in the environment and from direct contamination on crops and in milk. In the longer term, external radiation from radioactive isotopes of caesium and ruthenium is likely to become dominant, together with longer term contamination of foodstuffs with these radionuclides. Overall, during the first year following the accident, if no protective measures are taken, the largest component of the projected dose is likely to be received from contaminated foods, followed by external irradiation from contamination in the environment, with the smallest components originating from inhalation of radionuclides during dispersion of the plume or of resuspended radionuclides and external irradiation from the plume.

(A2) Assessments of the contributions of different exposure pathways to the projected dose can be based on numerical calculations using state-of-the-art radioecological models, which are readily available as part of decision support systems (Ehrhardt 1997, Ehrhardt and Weiss, 2000). The calculations require definition of a large number of input parameters. The most important are the characteristics of a release (total activity, nuclide vector, release height, and duration), the characteristics of the release site (urban/rural, flat terrain/complex topography), the time of year of a release (summer/winter), the meteorological conditions (wind strength and direction, atmospheric stability), the distance between a release and the area where people need protection, the human dietary and consumption rates, etc.

(A3) The Task Group has performed a variety of calculations of this type using standardised input parameters in order to demonstrate the application of the new system. Unit source terms for some key radionuclides, as well as complex source terms with a multitude of radionuclides which are typically used in risk assessment studies for nuclear power plants, have been used for this purpose.

(A4) Tables A.1 and A.2 (BMBF, 1990) show the results for the radionuclides Co-60, Sr-90, I-131, Cs-137, and Pu-239, as well as for a 'small release' (Sm_rel; nuclide vector which is characterised by the following cumulative fractions of the core inventory: Kr-Xe: 0.9, I: 2E-3, Cs: 3E-7, Te: 4E-6, Sr: 2E-7, Ru: 6E-10, La: 6E-8) and a 'large release' (Lg_rel; nuclide vector which is characterised by the following cumulative fractions of the core inventory: Kr-Xe: 1, I_{org}: 7E-3, I_2-Br: 4E-1, Cs-Rb: 2.9E-1, Te-Sb: 1.9E-1, Ba-Sr: 32.E-2, Ru: 1.7E-2, La: 2.6E-3). It is evident from the results of Tables A.1 and A.2 that there are major differences in the contributions of the pathways considered (ingestion, inhalation, cloud shine,

Table A.1. Characteristics in 'summer' (release 1 July).

Pathway	Co-60	Sr-90	I-131	Cs-137	Pu-239	Sm_rel	Lg_rel
Ingestion							
1 year	0.26	4.23	0.00	8.79	0.08	0.00	23.28
3 months	2.81	34.54	19.69	36.54	1.01	22.96	25.93
10 days	1.47	21.43	71.55	16.85	0.52	46.58	49.03
Inhalation	7.12	36.85	5.39	8.95	98.38	6.59	1.05
Ground							
1 year	63.72	2.14	0.00	21.00	0.00	0.00	0.09
3 months	20.43	0.69	1.10	6.73	0.00	0.08	0.11
10 days	3.83	0.13	2.21	1.02	0.00	0.38	0.37
Cloud	0.37	0.00	0.04	0.11	0.00	23.39	0.14
Total	100.00	100.00	100.00	100.00	100.00	100.00	100.00

Sm_rel, small release; Lg_rel, large release.

Table A.2. Characteristics in 'winter' (release 1 December).

Pathway	Co-60	Sr-90	I-131	Cs-137	Pu-239	Sm_rel	Lg_rel
Ingestion							
1 year	0.02	1.61	0.00	3.95	0.00	0.00	1.48
3 months	0.00	0.00	0.00	0.00	0.00	3.37	21.78
10 days	0.00	0.00	0.00	0.00	0.00	7.74	42.46
Inhalation	8.18	91.41	79.82	23.43	100.00	19.25	20.57
Ground							
1 year	66.70	5.17	0.01	53.56	0.00	0.00	1.70
3 months	21.35	1.57	6.82	16.27	0.00	0.25	2.06
10 days	3.33	0.24	12.77	2.50	0.00	1.11	7.21
Cloud	0.42	0.00	0.59	0.28	0.00	68.28	2.75
Total	100.00	100.00	100.00	100.00	100.00	100.00	100.00

Sm_rel, small release; Lg_rel, large release.

ground shine) to the total dose, which for the purpose of this presentation is normalised (100%). The main parameters are the radionuclides, the time of year, and the integration time after a release. For example, for Pu-239, inhalation contributes 100% to the total dose ('winter'), compared with 1% for a large release ('summer'). The relative contributions to the normalised yearly dose change with time; e.g. for I-131, ingestion contributes approximately 72% to the total annual dose for the first 10 days after release and another 20% by the end of 3 months. Of course, the absolute values of the annual dose would be markedly different for unit releases of the radionuclides in Tables A.1 and A.2. These differences have to be considered in an appropriate fashion in defining reference levels and in developing protection strategies in the planning state.

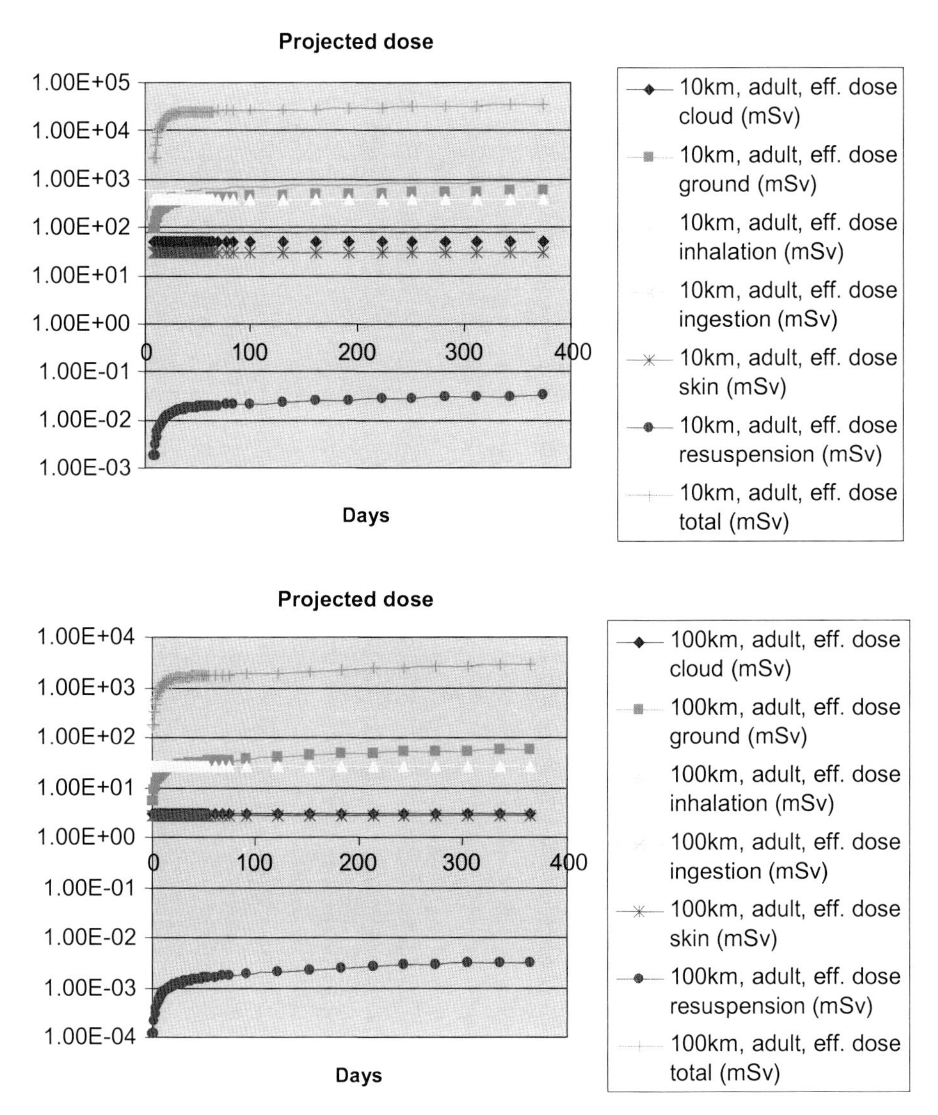

Fig. A.1. Time-dependent contributions of projected doses received via various exposure pathways to the total dose in a year.

(A5) The results presented in Tables A.1 and A.2 represent the conditions in the vicinity of a release. As a result of various processes (dilution due to atmospheric mixing and deposition on ground surfaces), the dose resulting from an airborne release will decrease with increasing distance between the point of release and the area for which protective measures have to be planned. Therefore, it is important to include this in the considerations during planning. Fig. A.1 demonstrates this effect for distances of 10 km (top) and 100 km (bottom).

A.1. References

BMBF, 1990. der Bundesminister für Forschung und Technologie (BMBF): Deutsche Risikostudie Kernkraftwerke, Phase B, Verlag TÜV Rheinland 1990, ISBN 3-88585-809-6.

Ehrhardt, J., 1997. The RODOS system: decision support for off-site emergency management in Europe. Radiat. Prot. Dosim. 731, 35–40.

Ehrhardt, J., Weiss, A., 2000. RODOS: Decision Support for Off-site Nuclear Emergency Management in EuropeEUR 19144 EN. European Community, Luxembourg.

ANNEX B. CHARACTERISTICS OF SELECTED INDIVIDUAL URGENT PROTECTIVE MEASURES

B.1. Iodine thyroid blocking

(B1) Iodine thyroid blocking is based on the administration of a compound of stable iodine (usually potassium iodide) to prevent or reduce the uptake of radioactive isotopes of iodine by the thyroid in the event of an accident involving radioactive iodine. Stable iodine is only of benefit in protecting the thyroid against radioactive iodine (reactor emergencies resulting in the release of radioactive iodine, laboratory emergencies, malicious events involving radioiodine).

(B2) Thyroid blocking prevents dose to the gland in case of exposure by inhalation and ingestion of radioiodines. However, as there is another measure that prevents radioiodine intake directly (restriction of potentially contaminated food consumption), thyroid blocking is considered to be primarily used for reduction of doses that result from inhalation. Iodine thyroid blocking should only be used to reduce the uptake of ingested radioiodine if it is impossible to provide supplies of uncontaminated food, especially for children and particularly in relation to milk; even if this is the case, iodine thyroid blocking is intended for relatively short periods of time, since efforts should be made to provide supplies of uncontaminated food as soon as possible. As iodine thyroid blocking is intended primarily as a protective measure against inhalation, it is therefore primarily a short-term measure (up to a few days) to reduce the risk of thyroid disease.

(B3) To obtain the maximum reduction of the radiation dose to the thyroid, stable iodine should be administered before any intake of radioiodine, or as soon as practicable thereafter. If stable iodine is administered orally within 6 h preceding the intake of radioactive iodine, the protection provided is almost complete; if stable iodine is administered at the time of radioiodine inhalation, the effectiveness of thyroid blocking is approximately 90%. The effectiveness of the measure decreases with delay, but the uptake of radioiodine can be reduced to approximately 50% if blocking is carried out within a few hours of inhalation.

B.2. Sheltering

(B4) Sheltering is the use of the structure of a building to reduce exposure from an airborne plume and/or deposited materials. Solidly constructed buildings can attenuate radiation from radioactive materials deposited on the ground and reduce exposure to airborne plumes. Buildings constructed of wood or metal are not generally suitable for use as protective shelters against external radiation, and buildings that cannot be made substantially airtight are not effective in protecting against any exposures.

(B5) Sheltering is not recommended for longer than approximately 2 days (IAEA, 1994, 1996, 2006). Sheltering is easy to implement but, in most cases, cannot be carried out for long periods. In addition, sheltering can be used as a preparation for an evacuation. The people in an area of potential risk can be instructed to 'go inside' and listen to their radios for further instruction while preparations for evacuation are being made. However, for very severe reactor accidents, sheltering in a typical home may not be sufficient to prevent deterministic health effects close to the facility. Sheltering is not a long-term protective measure; therefore, monitoring must be performed promptly anywhere sheltering is used in order to locate and evacuate people from areas of high risk.

B.3. Evacuation

(B6) Evacuation represents the rapid, temporary removal of people from an area to avoid or reduce short-term radiation exposure in an emergency exposure situation. It is most effective in terms of avoiding radiation exposure if it can be taken as a precautionary measure before there is any significant release of radioactive material. Generally, evacuation is not recommended for a period of longer than 1 week (IAEA, 1994, 1996, 2006).

B.4. Individual decontamination and medical intervention

(B7) Individual decontamination is the complete or partial removal of contamination from a person by a deliberate physical, chemical, or biological process.

(B8) Urgent individual decontamination may be advised to reduce exposures to external radiation from contamination on skin or inadvertent ingestion of such contamination. This measure may be particularly useful for protecting emergency workers. It is unlikely that individual decontamination will be required outside the area in which evacuation has been advised.

B.5. Preventative agricultural actions

(B9) Protective measures related to food can reduce or prevent doses from ingestion and include: a ban on the consumption of locally grown food in the affected area; the protection of local food and water supplies by, for example, covering open wells and sheltering animals and animal feed; and the long-term sampling and control of locally grown food and feed. Control of milk is very important because it is a significant part of the diet of children in many countries, as well as concentrating important radionuclides, such as radioiodine.

(B10) If appropriate, emergency plans should provide for considering the need for restriction of food consumption. Where they are needed, the population should be instructed not to drink milk from cows or goats that have been grazing on potentially contaminated pasture. In addition, they should be instructed not to eat fresh vegetables, fruit, or other food that may have been outside during the release and thereby contaminated. Drinking water is not typically a major concern during the initial response as it comes directly from collected rain water. However, water supplies should be monitored regularly during the response in case of a gradual build up of contamination owing to run-off into the water catchment area. If implemented, food and water restrictions should continue until sampling determines that food or milk is not contaminated beyond established levels.

B.6. References

IAEA, 1994. Intervention Criteria in a Nuclear or Radiation Emergency. Safety Series No. 109. International Atomic Energy Agency, Vienna.

IAEA, 1996. International Basic Safety Standards for Protection Against Ionizing Radiation and for the Safety of Radiation Sources. Safety Series No. 115. International Atomic Energy Agency, Vienna.

IAEA, 2006. Arrangements for Preparedness for a Nuclear or Radiological Emergency. IAEA Safety Standards Series No. GS-G-2.1. International Atomic Energy Agency, Vienna.

ANNEX C. SPECIFIC GUIDANCE FOR THE TERMINATION OF PROTECTIVE MEASURES

(C1) Although some protective measures may remain in effect for long periods, almost all protective measures will eventually need to be terminated. Some protective measures, particularly some of those that are implemented at an early stage, will, because of their nature (e.g. sheltering and use of stable iodine), be implemented for rather short periods (e.g. a one-off thyroid blocking with stable iodine, or sheltering implemented for few hours to a few days). Other measures, such as restrictions on the entry of food into the food chain, may last for longer periods of time. However, there are substantial risks associated with precipitate decisions to terminate protective measures before all the specific circumstances of the situation have been evaluated. For example, if protective measures are terminated too early, the situation may worsen unexpectedly, resulting in further exposures. The need for evaluation of the specific circumstances of the emergency and potential future exposures before making decisions on lifting protective measures means that it is difficult (or even potentially dangerous) to try to plan specific numerical guidance for this in advance. It is recognised that, to date, little guidance has been published on this topic. Therefore, the European Commission has developed a framework to assist the making of such decisions in real time. This annex discusses the Commission's guidance. In addition, emergency response experts in Europe have been addressing this topic. A recent report on their initial conclusions has been published as part of the European 'EURANOS' project (Nisbet et al., 2008).

(C2) The key problem for the decision makers is to balance the need to terminate unnecessary restrictions on people's lives with the need to ensure that the termination of protective measures does not expose them to unintended risks. This balance may vary between different population groups, and will be associated with considerable uncertainty. It may therefore be appropriate to treat population groups differently. Moreover, protective measures may be terminated at different times in different locations, as the required information on which to base the decision becomes available.

(C3) It may be appropriate to terminate protective measures for some groups of people, whilst continuing to recommend that they be left in place for other groups. This may be required because of local 'hotspots' or due to the inhomogeneity of the detailed monitoring. Whilst there may be clear radiological justification for this approach to lift protective measures, the potential for increased anxiety and misunderstanding of the revised advice needs to be recognised and addressed.

(C4) Owing to the passage of time and the possibility of scaling down the response, it is important for the decision makers to consider whether to replace existing protective measures with alternative measures. A prime example of this is sheltering. Sheltering cannot be maintained for a long period of time. From a radiological protection perspective, the protection afforded by sheltering will decrease with time, as radionuclides increasingly penetrate from the outside into the building. People require access to food, medicines, exercise, and contact with other people. Therefore, at some point, it will be necessary to discontinue sheltering, regardless of whether or not the release has stopped. At this point in time, the decision makers have to consider whether to terminate the advice to shelter altogether, or whether to replace it with alternative advice, such as evacuation.

(C5) Whilst every accident will have specific characteristics that influence decisions on how and when protective measures should be withdrawn, it is possible to provide general guidance on issues that require consideration and evaluation before such decisions are made. The following sections provide this guidance for the termination of protective measures implemented at an early stage, and the termination of those implemented at a later stage.

C.1. Guidance on the termination of urgent protective measures

(C6) The prime protective measures likely to be considered initially are the administration of stable iodine, advice to shelter or evacuate, and advice to avoid foods that may have become contaminated until a measurement programme can provide the information necessary to give more detailed advice regarding protective measures for food. The factors that require consideration for the termination of these protective measures differ between termination at an early stage and termination at a later stage after an accident.

(C7) There will be no requirement to consider the termination of evacuation or initial food advice until after the release has stopped. In the case of stable iodine, the decision is whether to advise a second administration of stable iodine in the event of a release lasting for more than 1 day, rather than whether or not to terminate this protective measure. In this situation, the Commission advises that if the population would otherwise require a second administration, every effort should be made to

Table C.1. Checklist for terminating the advice to shelter.

Issue	Comments/considerations
Duration	Unlikely to be practicable for more than 1 day.
Release status	Partial lifting (e.g. re-uniting of families) or phased evacuation may be considered before formal advice is given that the release has been terminated.
Contamination	Detailed monitoring in the sheltered area is likely to be a priority. Ensure that measurements are 'published' for access by media and public.
Information	Withdrawal of advice to shelter will probably be carried out without significant interaction with stakeholders because of the short time scales involved.
Health	Detailed information for all those affected is required for subsequent dose estimation and decisions on health follow-up programmes.
Stakeholders	Those affected should have been given the opportunity to contribute to the development of a protection strategy for withdrawing sheltering. They may need a mechanism to provide an input into decisions on the recovery strategy, should one be required.
Order of priority	Decisions on withdrawing sheltering will normally be accorded the highest priority.

Table C.2. Checklist for terminating the advice to evacuate.

Issue	Comments/considerations
Duration	Prolonged evacuation requires the provision of acceptable living conditions; many evacuation centres cannot provide such conditions. Supervised visits to the evacuated area in order to retrieve personal belongings or deal with animals left behind may reduce pressure for an early lifting of evacuation.
Release status	If a release occurs, the need to delay decisions on lifting of evacuation until a formal statement can be given that the release has definitely been terminated means that emergency plans should assume that evacuation will last from several days up to perhaps 1 week.
Contamination	Priority should be given to monitoring those areas in which it is expected that withdrawal of evacuation can take place. Measurement results should be 'published' for access by the media and the public.
Information	It is important to establish a mechanism for direct information and dialogue with the evacuees prior to their return to the area.
Health	Detailed information for all those affected is required for subsequent dose estimation and decisions on health follow-up programmes.
Stakeholders	Those affected should have been given the opportunity to contribute to the development of a protection strategy for stable iodine blocking. They may need a mechanism to provide input into decisions on the recovery strategy, should one be required.
Order of priority	Lower priority (relative to decisions on withdrawal of sheltering).

Note: it is important to distinguish between the admission of closely supervised people into an evacuated area in order to monitor, retrieve items, undertake maintenance activities, or provide security, and the provision of advice to an evacuated population to return home. This table provides a checklist for consideration before full withdrawal of evacuation.

Table C.3. Checklist for deciding that further doses of stable iodine should not be administered.

Issue	Comments/considerations
Duration	One dose of stable iodine will provide protection for approximately 24 h. Normally, evacuation would be preferred to administration of a second dose. Where the potential for prolonged releases indicates that multiple administrations to a sheltering population may be required, the emergency plan should address how this will be achieved.
Release status	Multiple administrations should not be considered unless a release is actually detected more than 24 h after the first administration, and evacuation is not practicable.
Contamination	Ideally, stable iodine prophylaxis should not be used to provide protection against contamination of food. Wherever practicable, restrictions on food should be implemented to provide protection against intake by ingestion.
Information	Stable iodine prophylaxis should be combined with either sheltering or evacuation, and has the same associated needs for information provision.
Health	The thyroids of infants born shortly before or shortly after the accident should be monitored individually if either the child or the mother received stable iodine. Details of all those who received stable iodine should be recorded in case of subsequent health problems.
Stakeholders	Those affected need a mechanism to provide input into decisions on the recovery strategy.
Order of priority	The multiple administration of stable iodine is only relevant to a sheltering population, which would normally be afforded the highest priority.

Table C.4. Checklist for terminating initial advice to avoid foods that may have been contaminated.

Issue	Comments/considerations
Duration	The precautionary measure of avoiding potentially contaminated foods can generally be maintained for up to a few days. After this time, the economic costs and, for some people, dietary needs may start to become a major issue, and so the precautionary advice must either be terminated or legally enforced on the basis of a measurement programme.
Release status	It is not possible to advise termination until after the release has been terminated.
Contamination	Priority should be given to monitoring those areas in which it is expected that termination of the food advice can take place. Measurement results should be 'published' for access by the media and the public.
Information	It is important to establish a mechanism for providing information to farmers, domestic producers, and those gathering foods from the wild.
Health	Information on the health risks of eating foods containing residual contamination should be readily available to all consumers.
Stakeholders	Those affected need a mechanism to provide input into decisions on the recovery strategy.
Order of priority	Decisions on lifting restrictions on food should only be afforded high priority where alternative food supplies are unavailable.

evacuate the population until the cessation of the release. Sheltering, however, is different. It may be necessary to terminate sheltering after a short period of time, either because it is not possible to maintain the measure for an extended period of time, or because it is decided that the population should be evacuated. In these situations, it is particularly important to determine how exposures and public anxiety and confidence will be impacted by the decision. Such decisions should be based on an informed understanding of the needs and concerns of the people affected, ideally through dialogue with potentially affected population groups prior to the occurrence of any accident. Such dialogue will assist in managing the expectations of those affected, so that sheltering followed by evacuation is expected. It will also inform decisions concerning how long sheltering can be maintained by different population groups, and whether specific support measures, such as the provision of emergency supplies and re-uniting of families, are practical alternatives to the early termination of sheltering.

(C8) As the emergency situation is developing, there may be very strong pressure to terminate all protective measures. However, it is important to fully evaluate the options and their likely consequences, and to avoid making precipitate decisions. Decisions to withdraw sheltering, evacuation, and food advice will need to reflect the prevailing circumstances of the emergency situation being addressed. Premature decisions to withdraw protective measures before all of the specific circumstances of the situation have been evaluated may result in further exposures if the situation worsens unexpectedly. In general, protective measures implemented at an early stage will be withdrawn because they have achieved their desired effect, or their continued application will cause more harm than good (e.g. sheltering for an extended period becomes too disruptive). Many different aspects must be taken into account when reaching such decisions, and as with all decisions regarding termination of protective measures, it is important to involve, wherever possible, relevant stakeholders in discussions. While it will be difficult, if not impossible, to discuss decisions with sheltered populations, it will be essential to discuss decisions to return to evacuated areas with those who have been evacuated. Non-radiological (e.g. economic, social, and psychological) consequences may become worse than the radiological consequences if there is a lack of pre-established guidance that is understandable to the public and officials. Tables C.1–C.4 summarise the key issues that should be considered.

(C9) One key difference between protective measures implemented at an early stage and those implemented at a later stage is that the former are likely to have been implemented on the basis of limited information concerning the actual situation and its impact. By the time when no further release is judged likely, additional information will have been gathered. This may demonstrate that the initial actions were an over-response, in which case there will be a strong incentive for those responsible for managing the response to reduce the extent and severity of protective measures as promptly as possible. However, even for these situations, it is important to ensure that the issues in Tables C.1–C.4 are explored in order to avoid the possibility of unexpected problems subsequently emerging.

C.2. Guidance on the termination of protective measures implemented in later stages

(C10) There is an important difference between protective measures initiated at an early stage and those implemented at later stages of an emergency exposure situation. The primary aim of urgent protective measures is to protect people from relatively high, short-term exposures. Decisions on their implementation are likely to be taken in a context of significant uncertainty. However, as the situation develops further, it will be much better characterised, whilst the protective measures put in place may well need to continue for weeks or months. These differences mean that for protective measures implemented at a later stage, it is both possible and desirable to establish the criteria for terminating those measures in advance of their initiation. These 'termination' criteria should be defined in terms of directly measurable or observable quantities, so that it is clear when they have been met. The criteria should also be discussed and agreed with stakeholders, so that the termination of the relevant protective measures is accepted. For some measures, e.g. permanent relocation, the criteria may be expressed in terms of the residual dose rates in the areas to be returned to; for others, such as decontamination, they are more likely to be expressed as the maximum level of residual contamination that will be accepted on the surface subject to a particular decontamination technique.

(C11) Protective measures implemented at later stages do not need to be initiated as urgently as those implemented at an early stage. This means that more time is available for dialogue with those who will be affected in order to implement the measures in a way that is truly optimal for stakeholders. Whilst it may not be possible to meet the expressed priorities of every individual, the process of involving stakeholders in decisions that are affecting their lives can help to reduce anxieties and frustrations, and contribute towards an efficient transition to management of the situation as an existing exposure situation.

C.3. Reference

Nisbet, A.F., Rochford, H., Cabianca, T., et al., 2008. Generic Guidance for Assisting in the Withdrawal of Emergency Countermeasures in Europe Following a Radiological Incident. EURANOS(CAT1)-TN(08)06. Available at: http://www.euranos.fzk.de/.

ELSEVIER

ICRP Publication 109

Annals of the ICRP

The History of ICRP and the Evolution of its Policies

R.H. Clarke and J.Valentin

Invited by the Commission in October 2008

Abstract–Within 12 months of the discovery of X rays in 1895, papers appeared in the literature reporting adverse effects from high exposure. In 1925, the first International Congress of Radiology, held in London, considered the need for a protection committee, which it established at its second congress in Stockholm in 1928. This paper celebrates the 80th anniversary of ICRP by tracing the history of the development of its policies, and identifying a few of the personalities involved from its inception up to the modern era. The paper follows the progress from the early controls on worker doses to avoid deterministic effects, through the identification of stochastic effects, to the concerns about public exposure and increasing stochastic risk estimates. The key features of the recommendations made by ICRP from 1928 up to the most recent in 2007 are identified.

Keywords: Occupational exposure; Public exposure; Medical exposure; Stochastic; Deterministic

1. INTRODUCTION

1.1. Sources

(1) This paper is based primarily on Clarke's (2008) presentation at the XII Congress of the International Radiation Protection Association (IRPA) and on a previous article by Clarke and Valentin (2005). It also draws extensively from Lindell (1996a). Additional sources include Lindell (1996b, 1999, 2003) and Taylor (1979). Valuable suggestions were provided by Clement (2009).

1.2. The discovery of radiation and its associated hazards

(2) Röntgen discovered X rays in November 1895 (Röntgen, 1895). Just a few months later, X-ray dermatitis was observed in the USA (Grubbé, 1933). Similar observations soon occurred in several countries; for instance, Drury (1896) described radiation damage to the hands and fingers of early UK experimental investigators, and Leppin (1896) made a similar report concerning German observations.

(3) Becquerel's (1896) identification of radioactivity, and the subsequent discovery of radium (Curie, 1898), led to many further cases of radiation damage, but the idea of inflicting such damage at will on selected tissues also paved the way for radiation therapy. The first proven cures of cancer patients were by Sjögren and Stenbeck in Sweden in 1899 (Mould, 1993).

(4) X rays were used by military field hospitals as early as 1897 (Churchill, 1898), although the number of X-ray injuries escalated during the Great War when primitive mobile X-ray equipment was used in the field. In the next 10 years, many papers were published on the tissue damage caused by radiation. However, during the first two decades following the discovery of X rays and radium, ignorance about the risks caused numerous injuries. Apparently, early radiologists often used their own hands to focus the beam of their X-ray machines, and skin cancer as a direct result of such exposure was described within 6 years of Röntgen's discovery (Frieben, 1902).

(5) The deleterious effects on hands and skin could be gruesome (as evidenced by the amputated hand of the German radiologist Professor Paul Krause at the Deutsches Röntgenmuseum in Remscheid). Unfortunately, it soon turned out that effects could be lethal, and the well-known monument to 'X-ray and radium martyrs' in Hamburg, erected in 1936 by the German Röntgen Society, names several hundred medical workers of many nationalities who died from radiation damage (Molineus et al., 1992).

1.3. The first protection recommendations

(6) Just 1 year after Röntgen's discovery of X rays, the American engineer Wolfram Fuchs (1896) gave what is generally recognised as the first protection advice. This was:

- make the exposure as short as possible;
- do not stand within 12 inches (30 cm) of the X-ray tube; and
- coat the skin with Vaseline (*a petroleum jelly*) and leave an extra layer on the most exposed area.

Thus, within 1 year of dealing with radiation, the three basic tenets of practical radiological protection – time, distance, and shielding – had been established!

(7) In the early 1920s, radiation protection regulations were prepared in several countries, but it was not until 1925 that the first International Congress of Radiology (ICR) took place and considered establishing international protection standards.

1.4. The International Commission on Radiological Protection

1.4.1. Its gestation and birth as IXRPC

(8) When the first ICR was held in London in 1925, the most pressing issue was that of quantifying measurements of radiation, and the International Commission on Radiation Units and Measurements (ICRU) was created, although it was then named the 'International X-ray Unit Committee'. The need for an international radiological protection committee was discussed, and the task was to ensure that a number of physicists interested in radiation protection would be present at the next ICR.

(9) The second ICR was held in Stockholm in 1928 and ICRU proposed the adoption of the röntgen unit; an event which was noted with far more interest than the birth of what is now ICRP under the name of the 'International X-ray and Radium Protection Committee' (IXRPC). As a courtesy to the host country, Rolf Sievert (who was then 32 years old) was named Chairman, but the driving person was George Kaye of the British National Physics Laboratory (Sievert, 1957; Lindell, 1996a). The other members present included Lauriston Taylor from the US National Bureau of Standards and Val Mayneord from the UK, who were in their 20s at the time. There were only two medical doctors on the Committee (Lindell, 1996a).

1.4.2. Development into maturity

(10) Before the Second World War, the Committee (or Commission, as it was called from 1934) was not active between the ICRs, and met for just 1 day at the ICRs in Paris in 1931, Zürich in 1934, and Chicago in 1937.

(11) Lindell (1996a) noted that at the 1934 meeting in Zürich, the Commission was faced with undue pressures; the hosts insisted on four Swiss participants (out of a total of 11), and the German authorities replaced the Jewish German member with another person. In response to these pressures, the Commission decided on new rules in order to establish full control over its future membership.

(12) After the Second World War, the first post-war ICR convened in London in 1950. Just two of the members of IXRPC had survived the war, namely Lauriston Taylor and Rolf Sievert. Taylor was invited to revive and revise the Commission,

which was now given its present name: the International Commission on Radiological Protection (ICRP). Sievert remained an active member, Sir Ernest Rock Carling (UK) was appointed as Chairman, and Taylor was Acting Secretary; after the ICR, Walter Binks (UK) took over as Scientific Secretary because of Taylor's concurrent involvement with the sister organisation, ICRU.

(13) At the 1950 meeting, a new set of rules was drafted, quite similar to the present rules, for the work of ICRP and the selection of its members (ICRP, 1951), and six sub-committees were established on:

- permissible dose for external radiation;
- permissible dose for internal radiation;
- protection against X rays generated at potentials up to 2 million volts;
- protection against X rays above 2 million volts, and β rays and γ rays;
- protection against heavy particles, including neutrons and protons; and
- disposal of radioactive wastes and handling of radioisotopes.

(14) It was also proposed (ICRP, 1951) that the Commission should 'recommend that all interested countries establish, each for itself, a central national committee to deal with problems of radiation protection – such a central committee to have sub-committees matching those of the International Commission on Radiological Protection as closely as their circumstances permit. So far as possible, members of the international sub-committees should be selected from the corresponding sub-committees of the various national committees. On matters of policy and formal agreements, communication will be from the central national committee to the International Commission. It is, however, recommended that direct communication on technical matters may be conducted between the corresponding national and international sub-committees'.

(15) This idea of a hierarchy of national and international committees and commissions, which never came into fruition, appears to herald Sievert's later visions of expanding ICRP into a single international authority, taking on the roles of the United Nations Scientific Committee on the Effects of Atomic Radiation (UNSCEAR) and other intergovernmental international organisations in radiation sciences and radiological protection (Lindell, 1999).

(16) However, it was now obvious that the amount of work expected from ICRP vastly exceeded what could be achieved by a handful of people meeting only in connection with the ICRs. An informal meeting was held at a radiobiology conference in Stockholm in 1952. The next formal meeting of ICRP took place at the seventh ICR in Copenhagen in 1953, and at that occasion, it was planned that the sub-committees proposed in 1950 would meet a week before the actual ICR started. However, since no member of the originally proposed Sub-Committee V on heavy particles was able to participate in Copenhagen, this was merged into Sub-Committee IV on other high-energy radiations. The original Sub-Committee VI failed to produce any report at the time, but was retained, now as Sub-Committee V. All sub-committee chairmen were members of the Commission. Thus, the structure used in 1953 was (ICRP, 1955):

SC I: Permissible doses for external radiation;
SC II: Permissible doses for internal radiation;
SC III: Protection against X rays generated at potentials up to three million volts;
SC IV: Protection against X rays above three million volts, β rays, γ rays, and heavy particles, including neutrons and protons; and
SC V: Handling and disposal of radioactive isotopes.

(17) The Commission and its committees (as they were now called) met again in the spring of 1956 in Geneva. This was the first time that a formal meeting of the Commission took place at a venue other than an ICR (the 1956 ICR was in Mexico, but ICRP simply could not afford to participate there). At this meeting, ICRP became formally affiliated with the World Health Organization (WHO) as a 'participating non-governmental organisation'. In 1959, a formal relationship had also been established with the International Atomic Energy Agency (IAEA), and various forms of relations were also in place with UNSCEAR, the International Labour Office (ILO), the Food and Agriculture Organization (FAO), the International Organization for Standardization (ISO), and the United Nations Educational, Scientific, and Cultural Organization.

(18) In the 1950s, the Commission did not have an administrative or financial basis commensurate with its increasing workload. Almost no funds were available to cover travelling costs, so members participated in meetings to the extent that they were able to obtain funding in their own countries (or if they could finance their participation with personal funds, as Sievert seems to have done on some occasions). Furthermore, until 1957, the Secretary of the Commission was simply one of the members who had accepted this as an honorary position. In 1957, the then Secretary, Walter Binks, had to retire for health reasons, and Bo Lindell (who was not a member of the Commission at the time) was first asked to be a 'temporary' secretary, then went on to become the Scientific Secretary, but still unpaid. Minor contributions had been received from the International Society of Radiology (ISR) from the National Association of Swedish Insurance Companies, and from 'private Swedish sources' (almost certainly, Sievert); some ad-hoc financing of meetings with UNSCEAR was also received directly from UNSCEAR.

(19) However, in 1960, the Ford Foundation provided a grant of $250,000 to ICRP, a very significant amount at the time, which meant that some funds were now available for secretariat and travelling and meeting costs. In the period 1960–1963, further grants totalling $60,000 were received from WHO, ISR, UNSCEAR, and IAEA, and the concept of grant applications and funding thus became established.

(20) The first full-time, paid Scientific Secretary of ICRP was F. David Sowby, who replaced Bo Lindell at the eighth ICR in Montreal, 1962. Sowby lived in Canada at the time, but moved to England where the new Chairman, Sir Edward ('Bill') Pochin, was based. The Secretariat was established at Sutton in Surrey, and the presence of a full-time employee greatly improved the efficiency of the work of ICRP.

1.4.3. 1962–2005: a modern structure

(21) At a meeting in Stockholm in May 1962, the Commission also decided to re-organise the committee system in order to improve productivity, which had differed considerably between committees. The five 'Roman numeral' committees were replaced with four committees with Arabic numerals, rather similar to the present Committees 1–4:

C1: Radiation effects;
C2: Internal exposure;
C3: External exposure; and
C4: Application of recommendations.

It was decided that the new committees should watch the development within various fields, and make suggestions on necessary actions to the Commission. Reports, however, should not be drafted by the committees but by ad-hoc task groups. The drafts would then be reviewed by the relevant committee before adoption by the Commission.

(22) However, even though the committees had been re-organised, several further reports were published under the banner of the earlier 'Roman numeral' committees, ending with *Publication 5* (ICRP, 1965) which was a report by Committee V on handling and disposal of radioactive isotopes.[1]

Defining the formal tasks of the Committees

(23) During the 1977–1981 term, the Commission reviewed and updated the names and missions of its committees. Increasing attention to organisational formalities can be seen in the fact that beginning in 1977, mission statements for each committee were published in leaflets describing ICRP. In the 1977 leaflet, Committee 1 retained (as it still does) its name, 'Radiation effects'. Its mission was given as follows: 'Committee 1 will assess the risk and severity of stochastic effects and the induction rates of the non-stochastic effects of irradiation. It will consider the modifying influence of exposure parameters such as dose rate, fractionation of dose, RBE, spatial distribution of dose and any synergistic effects of chemical and physical factors'. From 1981, the tense was amended: 'Committee 1 assesses...', 'It considers...' but otherwise the description remained unchanged until 1998 (see below).

(24) However, Committee 2 was no longer named 'Internal exposure'; in 1977, its name was changed to 'Secondary limits' and its mission statement was: 'The basic function of Committee 2 is to develop values of secondary limits, based on the Commission's recommended dose-equivalent limits. For the immediate future the committee will be fully concerned with the preparation of secondary limits for internal irradiation; because of this, matters to do with the derivation of secondary limits for external irradiation will, for the time being, be considered by Committee 3'.

[1] The sponsoring committee is given a Roman numeral in some later reports, but these instances simply reflect inconsistent copy-editing; i.e. those reports were really sponsored by the new 'Arabic numeral' committees.

(25) In 1981, the word 'fully' disappeared. In 1985, the description was abbreviated: 'Committee 2 develops values of secondary limits for internal and external irradiation, based on the Commission's recommended primary limits on dose equivalent'. At that time, Committee 2 had essentially completed the huge *Publication 30* on intake limits for workers, in three parts and four supplements plus a separate index issue, i.e. eight separate books comprising 26 standard issues of *Annals of the ICRP* (ICRP, 1979a,b, 1980, 1981, 1982a,b,c,d). An addendum constituting a fourth and final part appeared as ICRP (1989). Thus, it was considered that the work of Committee 2 on internal radiation had advanced far enough to permit the Committee to include, as originally intended, external irradiation in its scope. The name and the 1985 description remained unchanged until 1998.

(26) The new name given to Committee 3 in 1977, 'Protection in medicine' (which is still valid today), reflected a significant re-orientation of priorities. The mission was stated in 1977 as follows: 'The Commission considers that its relationship to the International Congress of Radiology and its traditional contacts with the medical profession warrant the establishment of a committee specifically concerned with radiation protection in medicine. Matters requiring particular attention by the committee include protection of the patient in radiodiagnosis and radiotherapy

Table 1.1. Names and mission statements of the ICRP committees.

Committee number	Committee name	Mission statement
Committee 1	Radiation effects	Committee 1 considers the risk of induction of cancer and heritable disease (stochastic effects) together with the underlying mechanisms of radiation action; also, the risks, severity, and mechanisms of induction of tissue/organ damage and developmental defects (deterministic effects).
Committee 2	Doses from radiation exposure	Committee 2 is concerned with development of dose coefficients for the assessment of internal and external radiation exposure, development of reference biokinetic and dosimetric models, and reference data for workers and members of the public.
Committee 3	Protection in medicine	Committee 3 is concerned with protection of persons and unborn children when ionising radiation is used for medical diagnosis, therapy, or for biomedical research; also, assessment of the medical consequences of accidental exposures.
Committee 4	Application of the Commission's recommendations	Committee 4 is concerned with providing advice on the application of the recommended system of protection in all its facets for occupational and public exposure. It also acts as the major point of contact with other international organisations and professional societies concerned with protection against ionising radiation.
Committee 5	Protection of the environment	Committee 5 is concerned with radiological protection of the environment. It aims to ensure that the development and application of approaches to environmental protection are compatible with those for radiological protection of man, and with those for protection of the environment from other potential hazards.

and protection in nuclear medicine. Committee 3 will temporarily be concerned with the development of secondary standards for external radiation'.

(27) In 1985, the last sentence about secondary standards for external radiation was removed, since that task had now been assigned to Committee 2. Also, the penultimate sentence was amended as follows: '...include protection of the patient and worker in radiodiagnosis, radiotherapy and nuclear medicine'.

(28) In the 1977 revision, Committee 4 kept its old name, 'Application of the Commission's recommendations', which is still valid today, and its mission was stated as follows: 'Committee 4 will continue its role of providing advice on the Commission's system of dose limitation, and on protection of the worker and the public. The committee will also serve as a major point of contact with international organisations concerned with radiation protection'. From 1981, the tense was amended: 'The Committee provides advice on the application of the Commission's...', 'It also serves...', but otherwise the description remained unchanged until 1998.

(29) In 1998, the Commission re-reviewed the entire set of names and mission statements. The name of Committee 2 was changed again, from 'Secondary limits' to its present version, 'Doses from radiation exposure'. The mission statements of all of the committees were updated at the same time (see Table 1.1).

Table 1.2. The officers of the Commission and its committees.

Position	Term	Name
IXRPC/ICRP and Main Commission Chair	1928–1931	Rolf Sievert, Sweden
	1931–1937	René Ledoux-Lebard, France
	1937–1950	Lauriston S. Taylor, USA
	1950–1956	Sir Ernest Rock Carling, UK
	1956–1962	Rolf Sievert, Sweden
	1962–1969	Sir Edward Eric ('Bill') Pochin
	1969–1977	C. Gordon Stewart, Canada
	1977–1985	Bo Lindell, Sweden
	1985–1993	Dan J. Beninson, Argentina
	1993–2005	Roger H. Clarke, UK
	2005–2009	Lars-Erik Holm, Sweden
	2009–	Claire Cousins, UK
Scientific Secretary	1928	George W.C. Kaye, UK
	1934, 1937	Lauriston S. Taylor, USA
	1947–1950	Lauriston S. Taylor, USA
	1950–1955	Walter Binks, UK
	1956	Eric E. Smith, UK
	1957–1962	Bo Lindell, Sweden
	1962–1985	F. David Sowby, Canada
	1985–1987	Michael ('Mike') C. Thorne, UK
	1987–1997	Hylton Smith, UK
	1997–2008	Jack Valentin, Sweden
	2009–	Christopher H. Clement, Canada

Table 1.3. The officers of the committees of ICRP.

Position	Term	Name
SC I / C I Chair	1950–1959	Giaocchino Failla, USA
	1959–1962	John F. Loutit, UK
SC II / C II Chair	1950–1962	Karl Z. Morgan, USA
SC III / C III Chair	1950–1962	Robert G. Jaeger, Austria
SC IV / C IV Chair	1950–1956	W. Valentine Mayneord, UK
	1956–1959	Harold E. Johns, Canada
	1959–1962	Gerard James Neary, UK
SC V / C V Chair	1950–1953	Dean B. Cowie, USA
	1953–1956	André J. Cipriani, Canada
	1956–1962	Conrad P. Straub, USA
SC VI Chair	1950–1953	Herbert M. Parker, UK
Committee 1 Chair	1962–1965	John F. Loutit, UK
	1965–1973	Howard B. Newcombe, Canada
	1973–1981	Arthur C. Upton, USA
	1981–1985	Dan J. Beninson, Argentina
	1985–2001	Warren K. Sinclair, USA
	2001–2009	Roger Cox, UK
	2009–	Ohtsura Niwa, Japan
Committee 2 Chair	1962–1973	Karl Z. Morgan, USA
	1973–1985	Jack Vennart, UK
	1985–1993	Charles ('Charlie') B. Meinhold, USA
	1993–2001	Alexander Kaul, Germany
	2001–2007	Christian Streffer, Germany
	2007–	Hans-Georg Menzel, Switzerland
Committee 3 Chair	1962–1965	Eric E. Smith, UK
	1965–1977	Bo Lindell, Sweden
	1977–1985	Charles B. Meinhold, USA
	1985–1993	Julian Liniecki, Poland
	1993–1996	Henri Jammet, France
	1996–2005	Fred J. Mettler, USA
	2005–2009	Claire Cousins, UK
	2009–	Eliseo Vañó, Spain
Committee 4 Chair	1962–1985	Henri Jammet, France
	1985–1989	H. John Dunster, UK
	1989–1993	Roger H. Clarke, UK
	1993–1997	Dan J. Beninson, Argentina
	1997–2003	Bert Winkler, South Africa
	2003–2009	Annie Sugier, France
	2009–	Jacques Lochard, France
Committee 5 Chair	2005–	R. Jan Pentreath, UK

1.4.4. 2005 and on: widening the scope beyond mankind

(30) In 2003, the Commission decided to launch a fifth committee, devoted to environmental protection, at the start of the 2005–2009 term, and the name and mission statement of Committee 5 were formally decided in 2004.

(31) The names and mission statements of the present five committees are given in Table 1.1. The Chairs and Scientific Secretaries of the Commission are listed in

Table 1.2, and the Chairs of the Sub-Committees and Committees are given in Table 1.3. The portrait annex at the end of this paper includes pictures of the Commission's Chairs and Scientific Secretaries, and some other key personalities.

1.5. References

Becquerel, H., 1896. Emission des radiations nouvelles par l'uranium metallique. C. R. Acad. Sci. Paris 122, 1086.

Churchill, W.S., 1898. The Story of the Malakind Field Force. Longman's Green & Co., London.

Clarke, R.H., 2008. The International Commission on Radiological Protection 80th Anniversary: the evolution of its policies through 80 years. In: Gallego, E., Pérez, M., Beatriz, G., et al. (Eds.), Strengthening Radiation Protection Worldwide, IRPA XII Congress. Available at: http://www.irpa12.org.ar/index.php.

Clarke, R.H., Valentin, J., 2005. A history of the International Commission on Radiological Protection. Health Physics 88, 407–422.

Clement, C., 2009. Personal communication.

Curie, M., 1898. Rayons emis par les composes de l'uranium et du thorium. C. R. Acad. Sci. Paris 126, 1101.

Drury, H.C., 1896. Dermatitis caused by Roentgen X-rays. Br. J. Med. 2, 1377.

Frieben, A., 1902. Demonstration eines Cancroid des rechten Handrückes, das sich nach langdauernder Einwirkung von Röntgenstrahlen entwickelt hat. Fortschr. Röntgenstr. 6, 106–111.

Fuchs, W., 1896. Simple recommendations on how to avoid radiation harm. Western Electrician 12.

Grubbé, E.H., 1933. Priority in the therapeutic use of X-rays. Radiology XXI, 156–162.

ICRP, 1951. International recommendations on radiological protection. Br. J. Radiol. 24, 46–53.

ICRP, 1955. Recommendations of the ICRP. Br. J. Radiol. (Suppl. 6) 100 pp.

ICRP, 1979a. Limits for intakes of radionuclides by workers. ICRP Publication 30, Part 1. Ann. ICRP 2(3/4).

ICRP, 1979b. Limits for intakes of radionuclides by workers. ICRP Publication 30, Supplement to Part 1. Ann. ICRP 3(1–4).

ICRP, 1980. Limits for intakes of radionuclides by workers. ICRP Publication 30, Part 2. Ann. ICRP 4(3/4).

ICRP, 1981. Limits for intakes of radionuclides by workers. ICRP Publication 30, Supplement to Part 2. Ann. ICRP 5(1–6).

ICRP, 1982a. Limits for intakes of radionuclides by workers. ICRP Publication 30, Part 3. Ann. ICRP 6(2/3).

ICRP, 1982b. Limits for intakes of radionuclides by workers. ICRP Publication 30, Supplement A to Part 3. Ann. ICRP 7(1–3).

ICRP, 1982c. Limits for intakes of radionuclides by workers. ICRP Publication 30, Supplement B to Part 3. Ann. ICRP 8(1–3).

ICRP, 1982d. Limits for intakes of radionuclides by workers. ICRP Publication 30, Index. Ann. ICRP 8(4).

ICRP, 1989. Limits for intakes of radionuclides by workers. ICRP Publication 30, Part 4 (Addendum). Ann. ICRP 19(4).

Leppin, O., 1896. Aus kleine Mitteilungen. Wirkung der Röntgenstrahlen auf die Haut. Dtsch. Med. Wschr. 28, 454.

Lindell, B., 1996a. The history of radiation protection. Rad. Prot. Dosim. 68, 83–95.

Lindell, B., 1996b. Pandoras Ask. Atlantis, Stockholm. (In Swedish; German translation 2004, Pandoras Büchse, Aschenbeck & Isensee, Oldenburg).

Lindell, B., 1999. Damokles Svärd. Atlantis, Stockholm. (In Swedish; German translation 2006, Das Damoklesschwert. Aschenbeck & Isensee, Oldenburg).

Lindell, B., 2003. Herkules Storverk. Atlantis, Stockholm. (In Swedish; German translation in preparation, Die Heldentaten des Herkules, Aschenbeck & Isensee, Oldenburg).

Molineus, W., Holthusen, H., Meyer, H., 1992. Ehrenbuch der Radiologen aller Nationen, third ed. Blackwell Wissenschaft, Berlin.
Mould, R.F., 1993. A Century of X Rays and Radioactivity in Medicine, second ed. CRC Press/Taylor & Francis Group, London.
Sievert, R.M., 1957. The International Commission on Radiological Protection (ICRP). In: International Associations. Union of International Associations, Palais d'Egmont, Brussels, pp. 3–7.
Röntgen, W.C., 1895. Über eine neue Art von Strahlen. Sitzungsberichte d. Phys. Mediz. Ges. Würzburg 9, 132.
Taylor, L.S., 1979. Organization for Radiation Protection: the Operations of the ICRP and NCRP 1928–1974. DoE/TIC 10124. US Department of Energy, Washington, DC.

2. HOW ICRP RECOMMENDATIONS EVOLVED

(32) The early recommendations of IXRPC were concerned with avoiding threshold (deterministic) effects, initially in a qualitative manner.

2.1. The initial stage: physical protection

(33) The Committee issued its first recommendations (IXRPC, 1928), consisting of 41 paragraphs, in three and a half pages of recommendations on protection against X rays and radium.

(34) The 1928 Recommendations noted that 'The effects to be guarded against are injuries to superficial tissues, derangements of internal organs and changes in the blood'. As a remedy, a prolonged holiday and limitation of working hours were recommended. No form of dose limit was proposed, but Lindell (1998) estimated that occupational annual effective doses to medical staff at the time may have averaged around 1000 mSv. The main emphasis of the 1928 Recommendations was of a technical nature on shielding requirements. There was also some practical guidance on protection in Paragraphs 10 and 11:

'(10) An X-ray operator should on no account expose himself unnecessarily to a direct beam of X-rays.

(11) An operator should place himself as remote as practicable from the X-ray tube. It should not be possible for a well rested eye of normal acuity to detect in the dark appreciable fluorescence of a screen placed in the permanent position of the operator.'

2.2. The first quantitative recommendations: tolerance dose

(35) IXRPC met again at the 1931 ICR in Paris but did not issue any recommendations at the time. However, at the next meeting, in Zürich, the first set of recommendations including a 'dose limit' (a limit on exposure rate for X rays) were issued. It was still stated that the 'known effects to be guarded against [were] injuries to the superficial tissues [and] derangements of internal organs and changes in the blood' (IXRPC, 1934).

(36) Furthermore, implying the concept of a safe threshold below which no untoward effects were expected, the 1934 Recommendations claimed that 'the evidence at present available appears to suggest that under satisfactory working conditions a person in normal health can tolerate exposure to X rays to an extent of about 0.2 international röntgens (r) per day. On the basis of continuous irradiation during a working day of seven hours, this figure corresponds to a dosage rate of 10^{-5} r per second. The protective values given in these recommendations are generally in harmony with this figure under average conditions'. This would correspond to an annual effective dose of approximately 500 mSv, i.e. approximately 25 times the

present annual limit for average occupational dose and approximately 10 times the present limit for occupational dose in any 1 year.

(37) In addition, the earlier (IXRPC, 1928) advice on working hours per day and per week, annual holiday, and freedom from 'other' hospital work were repeated in the 1934 Recommendations, and a requirement was added that 'X-ray, and particularly radium workers, should be systematically submitted, both on entry and subsequently at least twice a year, to expert medical, general and blood examinations. These examinations will determine the acceptance, refusal, limitation or termination of such occupation' (IXRPC, 1934).

(38) Interestingly, it was also stated that 'No similar tolerance dose is at present available in the case of radium gamma rays' (IXRPC, 1934). The general recommendations on shielding, electrical safety precautions (probably very important at the time), and working methods were extended. In addition, there was now advice on the safe storage of film, when non-flammable film was not available.

(39) The Commission met again at the next ICR meeting in Chicago in 1937. The amended recommendations issued at that time (IXRPC, 1938) were now for 'X rays and gamma rays', but did not differ from the 1934 Recommendations in any other respect than some extension of the practical guidance concerning shielding and electrical safety.

2.2.1. Radiological protection at the dawn of the nuclear age

(40) There were no further Commission recommendations before the Second World War. However, Sowby (1981) noted that the Commission's recommendations during the 1930s led to a great improvement in the standard of safety in radiological work. In particular, he stressed that the 'dose limits' introduced by ICRP served as the basis for the safety measures that were applied in the developing nuclear energy programmes. As a result, there were very few radiation injuries among the many thousands of workers involved in the early days of nuclear energy, despite the fact that large amounts of radioactive material were being handled.

2.3. A broader range of hazards: maximum permissible dose

(41) The first post-war meeting of the Commission, at the ICR in 1950, resulted in an 8-page report (ICRP, 1951). The Commission now recommended a maximum permissible dose of 0.5 röntgen in any 1 week in the case of whole-body exposure to x and gamma radiation (at the surface, corresponding to 0.3 röntgen in 'free air'), and 1.5 röntgen in any 1 week in the case of exposure of hands and forearms.

(42) In modern terminology, this corresponds approximately to an annual limit for occupational effective dose of 150 mSv (although the meaning and concept of a dose limit was different at the time). The previous limit of 1 röntgen/week (0.2 röntgen/day) was considered to be too 'close to the probable threshold for adverse effects'.

(43) The 1951 report of the Commission was quite comprehensive. There was a table of relative biological effectiveness (RBE) values and data on Standard Man.

Maximum permissible body burdens were given for 11 nuclides, including 0.1 μg for radium-226. It was recognised that in the case of uranium, it is the chemical toxicity and not radioactivity that is limiting. The 1950 Recommendations also provided an impressive list of the health effects that should be kept under review:

- superficial injuries;
- general effects on the body, particularly blood and blood-forming organs, e.g. production of anaemia and leukaemia;
- the induction of malignant tumours;
- other deleterious effects including cataract, [somewhat surprisingly] obesity, impaired fertility, and reduction of life span; and
- genetic effects.

(44) However, the casual reader might perceive the 1950 Recommendations to be somewhat inconsistent. On one hand, they mention 'permissible levels', 'maximum permissible exposure', and 'the probable threshold for adverse effects', all implying the existence of a safe threshold below which there would be no deleterious effects. On the other hand, they also mention the importance of carcinogenic and genetic effects (which, for the the latter effect at least, had been known for many years to operate at very low doses in experimental organisms), and it was 'strongly recommended that every effort be made to reduce exposures to all types of ionizing radiation to the lowest possible level'. The reduction of the 'dose limit' from approximately 500 mSv/year to approximately 150 mSv/year was not necessarily only due to the perception of 'new' biomedical hazards. It may also have reflected the realisation that there could be individual variations in radiation sensitivity. In any case, the 1950 Recommendations reflect a wide range of different opinions within the Commission at the time.

(45) In summary, for the first 60 years after the discovery of ionising radiation, the purpose of radiological protection was that of avoiding deterministic effects from occupational exposures, and the principle of radiological protection was to keep individuals below the relevant thresholds. The ethical basis of radiological protection was hardly discussed in any formal way, and seems to have been essentially a case of Aristotelian virtue ethics, i.e. having an 'inner sense' of moral orientation. Low doses of radiation were deemed beneficial, largely because most uses of radiation were for medical purposes, and radioactive consumer products abounded.

2.4. The need for change: public concern about radiation

(46) The next meeting of ICRP was at the seventh ICR in Copenhagen in 1953. The meeting generated a much more substantial set of recommendations, later referred to as the 1954 Recommendations (although they were agreed in 1953 and printed in 1955), comprising 100 pages (ICRP, 1955). The 1954 Recommendations included both the Commission's own recommendations and reports from Sub-Committees I–IV. The basic principle was re-iterated: 'Whilst the values proposed for maximum permissible doses are such as to involve a risk which is small compared to the other hazards of life, nevertheless, in view of the incomplete evidence

on which the values are based, coupled with the knowledge that certain radiation effects are irreversible and cumulative, it is strongly recommended that every effort be made to reduce exposure to all types of ionizing radiation to the lowest possible level'. This seems to be the first time that the Commission tried to put radiation risks into some form of perspective by comparing them with risks due to other factors.

(47) In the mid-1950s, there was growing public concern about radiation risks. The 'atomic bombs' dropped over Hiroshima and Nagasaki in 1945; the extensive nuclear weapons testing after the Second World War, with considerably larger explosive yields and resulting radioactive contamination in the northern hemisphere; and incidents such as the contamination of the Japanese fishing boat, the *Lucky Dragon*, in 1954 (Lapp, 1958) all influenced public opinion significantly.

(48) This development was also of concern for ICRP. The Commission recognised the need to protect the general public in the case of increasing use of radioactive sources, and with nuclear energy expected to be an expanding industry. The major problem, based on experimental data, was believed to be hereditary harm, but the awareness of leukaemia among radiologists, and information about increased frequency of leukaemia among the survivors in Hiroshima and Nagasaki, also contributed to a decision to be cautious with regard to public exposures.

(49) Thus, a first recommendation on restrictions of exposures of members of the general public appeared in the Commission's part of the 1954 Recommendations: 'The Commission recommends that, in the case of the prolonged exposure of a large population, the maximum permissible levels should be reduced by a factor of ten below those accepted for occupational exposures'. This is somewhat unspecific in the sense that 'large' populations was not defined, but it heralded the factor of 10 reduction that would, for many years, apply to the difference between occupational and public exposures.

(50) In this context, it is perhaps worth re-iterating that the Commission has never argued that there would be any reason (such as knowing the hazard, and/or receiving a high salary including some form of risk premium) to permit 'more' radiation in occupational contexts. Instead, the Commission's view has been and is that 'less' radiation should be permitted for the general public. Initially, this position was taken perhaps primarily in view of possible genetic effects, and later on the grounds that the general public includes more sensitive persons such as children and those suffering from diseases.

(51) In the report of Sub-Committee I in the 1954 Recommendations, it was stated that 'since no radiation level higher than the natural background can be regarded as absolutely "safe", the problem is to choose a practical level that, in the light of present knowledge, involves a negligible risk'. However, the Commission had not rejected the possibility of a threshold for stochastic effects.

(52) The Commission realised that it would no longer be sufficient to express all exposure restrictions in röntgen units. The 1954 Recommendations (ICRP, 1955) contained a glossary defining absorbed dose and the corresponding unit, rad (=0.01 Gy in modern terms), as described by ICRU. Sub-Committee I also introduced a new RBE-weighted unit, the rem (=0.01 Sv in modern terms).

(53) The concept of critical organ was now introduced, and the recommended dose limit was related to the organs that were said to be critical in the case of whole-body exposure, i.e. the gonads and the blood-forming organs. The limit, expressed in the new unit, was given as 0.3 rem/week, i.e. still corresponding to an annual occupational effective dose of the order of 150 mSv.

(54) The report of Sub-Committee II included tables on maximum permissible concentrations (MPC) in air and water for occupational exposure to some 90 radionuclides. These MPC values were all based on a weekly dose of 0.3 rem to the organ that was critical in each case. Concerning public exposures, it stated that 'Following accepted practice in the industrial and public health field, and keeping in mind the uncertainties involved and the fact that in the future some of the values given in this report may be lowered, it is recommended, in the case of prolonged exposure of a large population, to reduce by a factor of 10 the permissible level for radioactive isotopes accepted for occupational exposures. The values given in this report are for occupational exposure and are understood to be additional to natural background'.

(55) Furthermore, the Sub-Committee II report claimed that 'Exposure for a lifetime at the maximum permissible values recommended in this report is not expected, in the light of present knowledge, to cause appreciable body damage'. Since Sub-Committee I had specified that '"appreciable bodily injury" means any bodily injury or effect that a person would regard as being objectionable and/or competent medical authorities would regard as being deleterious to the health and well-being of the individual', this seems to imply the existence of a 'safe threshold'. Nevertheless, Sub-Committee II also stated that 'The application of the safety factor of 10 will reduce the risks of genetic damage that are a consequence of a large average exposure to the population'; thus the genetic risk, for which there was little reason to presume any threshold, was the primary concern.

(56) The Sub-Committee III report included an interesting proposal: 'In view of the continually increasing medical and technical use of ionizing radiation, it is desirable to accumulate information regarding the doses received both by individuals and by the population as a whole. As far as the individual is concerned, the information could be obtained by the introduction of a certificate in which are recorded details of all radiation exposure (medical and occupational) received through life. Probably it is impracticable to introduce such a certificate at present, but it is recommended that all radiologists and dentists keep records of the doses given'. The concept of an individual certificate for all forms of exposure never caught on, but of course nowadays significant occupational exposures are registered and stored electronically.

2.4.1. The 1956/57 amendment: controlled areas; pregnant women

(57) At its 1956 meeting in Geneva, the Commission concluded that the 1954 Recommendations needed a major and essential revision, implying substantial limitations of the MPC levels recommended earlier. However, it became obvious that a complete revision could not be completed before the summer of 1958. In order to promulgate the major points, the Commission released a 3-page amendment (ICRP, 1957) with several interesting conclusions. Thus, 'A controlled area is one in which

the occupational exposure of personnel to radiation or radioactive material is under the supervision of a radiation safety officer', and 'For any person in any place outside of controlled areas, the maximum permissible levels of exposure are 10% of the occupational exposure levels'. In other words, personnel working outside a controlled area were to be given the same level of protection as members of the public.

(58) For the entire population, in view of the genetic dose, 'it is prudent to limit the dose of radiation received by gametes from all sources additional to the natural background to an amount of the order of the natural background in presently inhabited regions of the earth'.

(59) Furthermore, 'Since it is known from animal experiments that the embryo is very radiosensitive, special care should be exercised to make sure that pregnant women are not occupationally exposed under conditions in which, through some accident or otherwise, they may be exposed to large doses of penetrating radiation. When the exposure cannot exceed the basic permissible weekly dose, however, no special provisions need be made'. This was the Commission's first specific advice for pregnant women, and at the same time, the first observation that a steady dose rate is essential for the protection of the embryo and fetus.

2.5. The beginnings of the modern era: ICRP *Publication 1*

(60) In 1957, there was pressure on ICRP from both WHO and UNSCEAR to reveal all of the decisions from its 1956 meeting in Geneva. The final document, the Commission's 1958 Recommendations (ICRP, 1959a [adopted in September 1958]) was the first ICRP report published by Pergamon Press. Although it had no number, the next report (ICRP, 1959b) had '*Publication 2*' printed on the cover, and therefore the 1958 Recommendations are usually referred to as '*Publication 1*'.

(61) The 1958 Recommendations, 22 pages comprising 87 paragraphs, began with a 'Prefatory review' and was the first time that the basis of the Commission's policy was presented and discussed. Both the 1954 and the 1958 Recommendations include a paragraph headlined 'Policy', but this just refers to the administrative policy of dealing with the basic principles of radiological protection and leaving detailed technical regulations to national bodies. The protection policy review in the 1958 Recommendations is printed in the Prefatory review.

(62) The weekly dose limit of 0.3 rem was replaced by a limit of the accumulated dose equivalent, $D = 5\ (N\text{-}18)$ where D is dose in rems and N is age in years, corresponding to an average annual occupational effective dose of 5 rem (50 mSv). For individual members of the public, the dose limit was set at 0.5 rem (5 mSv)/year and, in addition, a genetic dose limit of 5 rem/generation was suggested together with a long and detailed 'illustrative apportionment'.

(63) At that time, the Commission's basic policy was mainly determined by Committee I. The 1958 Recommendations were soon supplemented with *Publication 2* (ICRP, 1959b), a major document on internal emitters and with comprehensive tables on maximum permissible body burdens and MPC values, and *Publication 3* (ICRP, 1960), on protection against X rays and β and γ rays from sealed sources.

Together, *Publications 1–3* definitely established ICRP as the leading international radiation protection authority.

(64) *Publication 1* was soon subject to a number of amendments, and the Commission therefore issued a revised version, *Publication 6* (ICRP, 1964) in which these amendments were incorporated. This also included new MPC values for strontium-90 and some isotopes of transuranic elements.

2.6. Taking stochastic effects into account: the linear, no-threshold model

(65) The significance of stochastic effects began to influence the Commission's policy more and more. It was soon time for more substantial revisions, and a new set of recommendations was published as *Publication 9* (ICRP, 1966b). During the drafting of *Publication 9*, its editorial group had been concerned about the many different opinions regarding the risk of stochastic effects. The Commission therefore asked a working group 'to consider the extent to which the magnitude of somatic and genetic risks associated with exposure to radiation can be evaluated'. Their report, *Publication 8* (ICRP, 1966a) was an important document because, for the first time in ICRP publications, it summarised the current knowledge about radiation risks, both somatic and genetic. The probability of leukaemia after an absorbed dose of 1 rad of gamma radiation (i.e. 10 mGy) was estimated at 20 cases per million exposed. However, it was then assumed that the probability of all other types of cancer together was about the same as the probability of leukaemia; an assumption that was later shown to have been an underestimate.

2.6.1. Acceptable risks

(66) Prolonged debate followed regarding how to deal with the acceptability of the risks. In *Publication 1*, the 1954 words 'lowest possible' were succeeded by 'as low as practicable'. In *Publication 9*, the usual cautious warning (in Paragraph 52) read: 'As any exposure may involve some degree of risk, the Commission recommends that any unnecessary exposure be avoided and that all doses be kept as low as is readily achievable, economic and social consequences being taken into account'.

(67) Other considerations, such as ethical issues, were not excluded by this wording, but the Commission considered them to be included in the adjective 'social'. No guidance existed regarding how this recommendation should be applied. However, the Commission was increasingly doubtful of the existence of a threshold dose for the induction of cancer. Paragraph 7 stated that 'the Commission sees no practical alternative, for the purposes of radiological protection, to assuming a linear relationship between dose and effect, and that doses act cumulatively. The Commission is aware that the assumptions of no threshold and of complete additivity of all doses may be incorrect, but is satisfied that they are unlikely to lead to the underestimation of risks'.

(68) Now there were stochastic effects, where the probability of the effect, not the severity, is proportional to the size of the dose, the assumption of a threshold was rejected. The problem had become one of limiting the probability of harm. Much

of the subsequent development related to the estimation of that probability of harm, and the decision on what level of implied risk is acceptable or, more importantly, unacceptable. From the mid-1960s, the main field of interest was the expanding nuclear industry. The protection philosophy was definitely shaped by the assumption of a linear dose–response relationship without any threshold dose.

(69) *Publication 9* substantially renewed the radiation protection philosophy by moving from deterministic to stochastic effects. It made a distinction between 'normal operations' and accidents where the exposure 'can be limited in amount only, if at all, by remedial action'. The age-prorated formula was abandoned and the MPC for the gonads and the blood-forming organs was now expressed as an annual dose of 5 rem (i.e. 50 mSv). The term 'dose limit' was introduced for the annual limit of 0.5 rem recommended for public exposures.

(70) Paragraph 52 in *Publication 9*, recommending that 'all doses be kept as low as is readily achievable, economic and social consequences being taken into account', called for further guidance. The Commission therefore appointed a task group, which reported in *Publication 22* (ICRP, 1973) that the optimum level of protection might be found by means of differential cost–benefit analysis and that the principle described in Paragraph 52 of *Publication 9* was the principle of optimisation of protection.

(71) At that time, ICRP had a new editorial group working on a revision of *Publication 9* and proposed some rather radical changes. The concept of 'critical organ' was abandoned. It was felt that there was sufficient knowledge of the cancer risk for a number of organs to permit the calculation of a weighted whole-body dose. A quantity based on such weighting had already been suggested by Jacobi (1975), but in the new recommendations, the Commission only introduced the weighting procedure without presenting the result as a new quantity. This was first made in a statement (ICRP, 1978), when the name 'effective dose equivalent' was introduced, following Wolfgang Jacobi's proposal.

2.7. A system of dose limitation

(72) *Publication 26* (ICRP, 1977), the Commission's 1977 Recommendations, first quantified the risks of stochastic effects of radiation and proposed a system of dose limitation. The 1977 Recommendations stated in Paragraph 6 that 'Radiation protection is concerned with the protection of individuals, their progeny and mankind as a whole, while still allowing necessary activities from which radiation exposure might result'. The 1977 Recommendations then went on to say in Paragraph 14 that 'Although the principal objective of radiation protection is the achievement and maintenance of appropriately safe conditions for activities involving human exposure, the level of safety required for the protection of all human individuals is thought likely to protect other species, although not necessarily individual members of those species. The Commission therefore believes that if man is adequately protected then other living things are also likely to be sufficiently protected'.

(73) This was the first occasion on which the Commission addressed the effects of radiation on species other than mankind, although clearly it was not pursued. Much of the work of ICRP was concentrated upon the development of human biokinetic data, and the assessment of doses for workers and the public from the ranges of radionuclides likely to be encountered. This included the development of Reference Man to develop standardised dose-intake data.

(74) *Publication 26* set out the new system of dose limitation and introduced the three principles of protection in Paragraph 12:

'(a) no practice shall be adopted unless its introduction produces a positive net benefit;
(b) all exposures shall be kept as low as reasonably achievable, economic and social factors being taken into account; and
(c) the doses to individuals shall not exceed the limits recommended for the appropriate circumstances by the Commission'.

These principles have since become known as justification, optimisation (as low as reasonably achievable), and the application of dose limits.

(75) The principle of optimisation was to generate much important work for ICRP, as well as other international and national bodies. The principle was introduced because of the need to find some way of balancing costs and benefits of the introduction of a source involving ionising radiation or radionuclides. This process was not necessarily sufficient to protect individuals, so it was complemented by the dose limits. As a result of introducing this requirement, doses to non-human species were certainly reduced to some extent in the majority of situations.

(76) The 1977 Recommendations were very concerned with the bases for deciding what is reasonably achievable in dose reduction. The principle of justification aims to do more good than harm, and that of optimisation aims to maximise the margin of good over harm for society as a whole. They therefore satisfy the utilitarian principle of ethics, also called 'consequence ethics', proposed primarily by Jeremy Bentham and John Stuart Mill (Mill, 2002). Utilitarians judge actions by their overall consequences, usually by comparing, in monetary terms, the relevant benefits (e.g. statistical estimates of lives saved) obtained by a particular protective measure with the net cost of introducing that measure.

(77) On the other hand, the principle of applying dose limits aims to protect the rights of the individual not to be exposed to an excessive level of harm, even if this could cause great problems for society at large. This principle therefore satisfies the deontological principle of ethics, also called 'duty ethics', proposed primarily by Immanuel Kant (Broad, 1978). Proponents of this principle emphasise the strictness of moral limits.

(78) Paragraph 72 of *Publication 26* suggests that the decision on what is 'as low as reasonably achievable' depends on the answer to the question 'whether or not the activity [under scrutiny] is being performed at a sufficiently low level of collective dose equivalent (and usually, therefore, of detriment) so that any further reduction in dose would not justify the incremental cost required to accomplish it'.

(79) Paragraph 75 of *Publication 26* recommended the use of differential cost–benefit analysis where the independent variable is the collective dose, and recommended that a monetary value should be assigned to a unit of collective dose. This classical use of cost–benefit analysis addresses the question: 'How much does it cost and how many lives are saved?' However, this approach does not allow for the protection of the individual from the source, so ICRP retained the concept of a dose limit to protect the individual from all sources under control.

(80) The concept of the collective dose was originally introduced for two reasons, one of which was to facilitate cost–benefit analysis. The second reason for using collective dose was to restrict the uncontrolled build-up of exposure to long-lived radionuclides in the environment. This was because, at the time, a global expansion of nuclear power reactors and reprocessing facilities was foreseen, and there were fears that global doses could again reach the levels seen from atmospheric testing of nuclear weapons. Restricting collective dose per unit of practice can effectively set a maximum future annual effective per caput dose from all sources from that practice.

(81) In 1977, the establishment of the dose limits was of secondary concern to the establishment of cost–benefit analysis and use of collective dose. This can be seen in the wording used in *Publication 26* in setting its dose limit for members of the public: 'The assumption of a total risk of the order of 10^{-2} Sv^{-1} would imply restriction of the lifetime dose to the individual member of the public to 1 mSv/year. The Commission's recommended limit of 5 mSv in a year, as applied to critical groups, has been found to give this degree of safety and the Commission recommends its continued use'.

(82) In a similar manner, the dose limit for workers was argued on a comparison of average doses, and therefore risk, in the workforce, with average risks in industries that would be recognised as being 'safe', and not on maximum risks to be accepted.

2.8. Acceptability vs risk: tolerable detriment

(83) During the 1980s, there were re-evaluations of the risk estimates derived from the survivors of the atomic bombing at Hiroshima and Nagasaki, partly due to revisions in the dosimetry. The risks of exposure were claimed to be higher than those used by ICRP, and pressures began to appear for a reduction in dose limits. This represented the start, as now seen with hindsight, of the rise of concern regarding the individual. ICRP's response was initially to emphasise the principle of optimisation and to claim that the use of collective dose and cost–benefit analysis always ensured that individual doses were sufficiently low.

(84) However, by 1989, the Commission had itself revised upwards its estimates of the risks of carcinogenesis from exposure to ionising radiation. The following year, it adopted its 1990 Recommendations (ICRP, 1991) for a 'system of radiological protection'. The principles of protection recommended by the Commission were still based on the general principles given in *Publication 26*, but with important additions:

'(a) No practice involving exposures to radiation should be adopted unless it produces sufficient benefit to the exposed individuals or to society to offset the radiation detriment it causes. (The justification of a practice);

(b) In relation to any particular source within a practice, the magnitude of individual doses, the number of people exposed, and the likelihood of incurring exposures where these are not certain to be received should all be kept as low as reasonably achievable, economic and social factors being taken into account. This procedure should be constrained by restrictions on the doses to individuals (dose constraints), or on the risks to individuals in the case of potential exposures (risk constraints) so as to limit the inequity likely to result from the inherent economic and social judgements. (The optimisation of protection);

(c) The exposure of individuals resulting from the combination of all the relevant practices should be subject to dose limits, or to some control of risk in the case of potential exposures. These are aimed at ensuring that no individual is exposed to radiation risks that are judged to be unacceptable from these practices in any normal circumstances. (Individual dose and risk limits)'.

(85) The most significant change was in the principle of optimisation and the introduction of the concept of a constraint. Optimisation is a source-related process, while limits apply to the individual to ensure protection from all sources under control. The aim of dose limitation is to ensure that no individual is exposed to an unacceptable level of risk from all the regulated sources. The constraint is an individual-related criterion, applied to a single source in order to ensure that the most exposed individuals are not subjected to undue risk from that source. Classical cost–benefit analysis is unable to take this into account, so the Commission established an added restriction on the optimisation process, the maximum individual dose (or risk= probability of exposure) from the source, i.e., the constraint.

(86) In *Publication 77* (ICRP, 1998), the Commission observed that 'The optimisation of protection has the broad interpretation of doing all that is reasonable to reduce doses. In some ways it is unfortunate that the shorthand label "optimisation of protection" lost the adjective "reasonable" in the phrase "as low as reasonably achievable". Furthermore, the perception of optimisation of protection has become too closely linked to differential cost–benefit analysis'. Furthermore, it was stated that 'The unlimited aggregation of collective dose over time and space into a single value is unhelpful because it deprives the decision maker of much necessary information. The levels of individual dose and the time distribution of collective dose may be significant factors in making decisions'.

(87) In other words, this report weakened the link to cost–benefit analysis and collective dose. Thus, concern for the protection of the individual was being strengthened. This was a reflection of changing societal values, with more concern about individual welfare.

2.9. Emphasising individual rights, widening the scope to all species

(88) Since *Publication 60*, a series of publications has provided additional guidance for the control of exposures from radiation sources. When the 1990 Recommendations are included, these reports specify some 30 different numerical values for

restrictions on individual dose for differing circumstances. Furthermore, these numerical values are justified in many different ways (ICRP, 2006). In addition, the Commission began to develop policy guidance for protection of non-human species in *Publication 91* (ICRP, 2003).

(89) The Commission was elaborating its policy but it was clear that there were some misunderstandings of its concepts, in particular, the difference between source-related and individual-related protection. The dose limit as defined in the 1990 Recommendations only applies in defined conditions, but many people regarded a limit as being absolute. The use of higher doses for emergencies and for radon in homes was seriously confusing. The Commission had tried to clarify this by distinguishing between 'practices' that added doses and 'interventions' that subtracted doses, but the distinction was not clearly understood.

(90) Other factors that caused concern included the excessive formality of the use of differential cost–benefit analysis and the rigid interpretation of collective dose by some practitioners. This led to the initiation of a wide-ranging open review of the basis for protection philosophy (Clarke, 1999).

(91) The Commission prepared *Publication 103*, its 2007 Recommendations (ICRP, 2007), after two phases of international public consultation on drafts, one in 2004 and one in 2006, as well as presentations to IRPA and other international bodies as the drafts were developed. This process follows nearly a decade of a policy of transparency and involvement of those with a serious interest in protection, which the Commission expects to lead to a clear understanding and wide acceptance of its 2007 Recommendations.

(92) There is, therefore, more continuity than change in the 2007 Recommendations. Some recommendations remain because they work and are clear, others have been updated because understanding has evolved, some items have been added because there has been a void, and some concepts are better explained because more guidance is needed. The 2007 Recommendations re-iterate and strengthen the importance of optimisation in radiological protection, and extend the successful experience in the implementation of this requirement for practices (now included in planned exposure situations) to other situations, i.e. emergency and existing exposure situations. They also include a commitment to environmental protection.

2.10. References

Broad, C.D., 1978. Kant: an Introduction. Cambridge University Press, Cambridge.

Clarke, R.H., 1999. Control of low-level radiation exposure: Time for a change? J. Radiol. Prot. 19, 107–115.

ICRP, 1951. International recommendations on radiological protection. Revised by the International Commission on Radiological Protection and the 6th International Congress of Radiology, London, 1950. Br. J. Radiol. 24, 46–53.

ICRP, 1955. Recommendations of the International Commission on Radiological Protection. Br. J. Radiol. (Suppl. 6) 100 pp.

ICRP, 1957. Reports on amendments during 1956 to the Recommendations of the International Commission on Radiological Protection (ICRP). Acta Radiol. 48, 493–495.

ICRP, 1959a. Recommendations of the International Commission on Radiological Protection. ICRP Publication 1. Pergamon Press, Oxford.
ICRP, 1959b. Recommendations of the International Commission on Radiological Protection. ICRP Publication 2: Report of Committee II on Permissible Dose for Internal Radiation. Pergamon Press, Oxford.
ICRP, 1960. Recommendations of the International Commission on Radiological Protection. ICRP Publication 3: Report of Committee III on Protection Against X-Rays up to Energies of 3 MeV and Beta- and Gamma-rays from Sealed Sources. Pergamon Press, Oxford.
ICRP, 1964. Recommendations of the International Commission on Radiological Protection. ICRP Publication 6. Pergamon Press, Oxford.
ICRP, 1966a. The Evaluation of Risks from Radiation; a Report of Committee 1 of ICRP. ICRP Publication 8. Pergamon Press, Oxford.
ICRP, 1966b. Recommendations of the International Commission on Radiological Protection. ICRP Publication 9. Pergamon Press, Oxford.
ICRP, 1973. Implications of Commission Recommendations that Doses be Kept as Low as Readily Achievable. ICRP Publication 22. Pergamon Press, Oxford.
ICRP, 1977. Recommendations of the International Commission on Radiological Protection. ICRP Publication 26. Ann. ICRP 1(3).
ICRP, 1978. Statement from the 1978 Stockholm Meeting of the ICRP. ICRP Publication 28. Ann. ICRP 2(1).
ICRP, 1991. 1990 Recommendations of the International Commission on Radiological Protection. ICRP Publication 60. Ann. ICRP 21(1–3).
ICRP, 1998. Radiological protection policy for the disposal of radioactive waste. ICRP Publication 77. Ann. ICRP 27(Suppl.).
ICRP, 2003. A framework for assessing the impact of ionising radiation on non-human species. ICRP Publication 91. Ann. ICRP 33(3).
ICRP, 2006. Analysis of the criteria used by ICRP to justify the setting of numerical values. ICRP Supporting Guidance 5. Ann. ICRP 36(4).
ICRP, 2007. The 2007 Recommendations of the International Commission on Radiological Protection. ICRP Publication 103. Ann. ICRP 37(2–4).
IXRPC, 1928. X-ray and Radium Protection. Recommendations of the 2nd International Congress of Radiology, 1928. Br. J. Radiol. 12, 359–363.
IXRPC, 1934. International Recommendations for X-ray and Radium Protection. Revised by the International X-ray and Radium Protection Commission and adopted by the 4th International Congress of Radiology, Zürich, July 1934. Br. J. Radiol. 7, 1–5.
IXRPC, 1938. International Recommendations for X-ray and Radium Protection. Revised by the International X-ray and Radium Protection Commission at the 5th International Congress of Radiology, Chicago, September 1937. Br. Inst. Radiol. (leaflet) 1–6.
Jacobi, W., 1975. The concept of the effective dose – a proposal for the combination of organ doses. Rad. Environm. Biophys. 12, 101–109.
Lapp, R.E., 1958. The Voyage of the Lucky Dragon. Harper & Bros., New York.
Lindell, B., 1998. Personal communication.
Mill, J.S., 2002. Utilitarianism and On Liberty. Including 'Essay on Bentham' and Selections from the Writings of Jeremy Bentham and John Austin. Warnock, M. (Ed.), second ed. Blackwell Publishing, Oxford.
Sowby, F.D., 1981. Radiation protection and the International Commission on Radiological Protection (ICRP). Radiat. Prot. Dosim. 1, 237–240.

3. ANALYSIS OF TRENDS AND CONCLUSION

3.1. The formal status of ICRP

(93) ICRP is an independent, international, non-governmental organisation. It remains one of three commissions of the International Society for Radiology, the others being ICRU and the International Commission for Radiological Education and Information (ICRE), and the parent body approves the rules by which the three commissions operate. Various international (but not national) organisations are invited to send observers to ICRP meetings, and ICRP has a corresponding observer status in such organisations.

(94) The 1934 meeting of IXRPC and the membership discussions at that time was the first, but certainly not the last, time that the Commission's independence and scientific integrity were jeopardised by demands from special interest groups and other outsiders with vested interests. It still happens from time to time that the Commission is subjected to demands or covert criticisms aimed at gaining outside control of its membership and/or its policies. The Commission is very wary of such attempts and maintains as its strict policy that, as for ICRU and ICRE, members are elected by the Commission itself. Outside nominations are accepted as a means to achieve the widest possible range of expertise, but the actual elections are made by the Commission alone, and solely on the grounds of scientific merit, not as representatives of any country, organisation, or other entity.

3.1.1. Science, policy, legislation, and the role of ICRP

(95) Fig. 3.1 shows the relationship between different organisations with an interest in radiological protection. ICRP uses the summaries of basic scientific studies provided by UNSCEAR (e.g. UNSCEAR, 2006) as a primary source of information; it also takes account of scientific developments reported by major national organisations (e.g. the Biologic Effects of Ionizing Radiation reports of the US National Academies). Such scientific studies and summaries answer the questions 'How much radiation is there?' and 'How dangerous is it?'.

(96) Using that information as an input, the Commission proposes protection policies aimed at both legislators and regulators, operators and licensees, and, ultimately, members of the public (e.g. ICRP, 2007). Initially, and well through the 1960s, some of the advice given was of a very practical nature. More recently, with an increased availability of expertise in health physics, radiobiology, and radiological protection, ICRP recommendations focus primarily on the strategy of protection – the 'whether' and 'why' rather than the 'how'.

(97) The transparency of ICRP operations and the interaction with society have increased significantly in recent years. For instance, since 2002, all draft ICRP reports are subjected to public consultation using the Internet. Professional bodies, such as IRPA and, of course, ISR, provide important input to the Commission in this process.

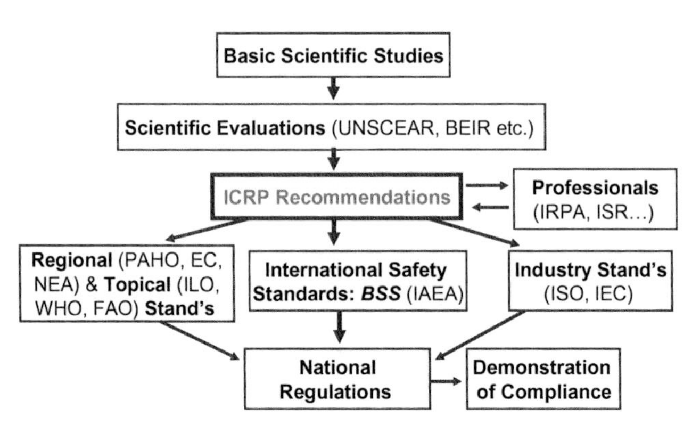

Fig. 3.1. The basis for and use of ICRP recommendations on radiological protection policy.

(98) The international organisations of the United Nations system utilise the recommendations of ICRP when producing their Basic Safety Standards for radiation protection (IAEA, 1996). The recommendations and advice of ICRP also influence documents on specific issues produced by specialised United Nations agencies, such as ILO, WHO, FAO, and IAEA. The Pan-American Health Organization, the European Commission, and the Nuclear Energy Agency of the Organisation for Economic Co-operation and Development (OECD) can be regarded as regional organisations (with OECD an economic region rather than a geographic region); they all take account of ICRP advice when producing documents pertaining to radiological protection. The International Electrotechnical Commission and ISO take ICRP advice into account when producing standards.

(99) Thus, neither the idea of a hierarchy of national and international committees and commissions (cf. Section 1.4.2) nor Rolf Sievert's vision of ICRP as a central intergovernmental international agency was implemented. However, with hindsight, the current system seems quite adequate. With UNSCEAR providing scientific summaries of levels and effects of radiation, ICRP providing policy recommendations, and the various intergovernmental agencies with an interest in radiological protection proposing regulations, purely scientific factors, political factors, and stakeholder demands are all given their due consideration but at separate and clearly identifiable stages.

3.2. Different aspects of the development of ICRP recommendations

(100) Table 3.1 summarises a number of different parameters characterising the Commission's recommendations over the years.

3.2.1. Exposure conditions considered

(101) Before the Second World War, the Commission's recommendations were entirely devoted to occupational exposures in medicine. Medical applications

Table 3.1. The historical development of ICRP recommendations.

Factor analysed	Early recommendations	Intermediate recommendations	Present recommendations
Circumstances of exposure considered	Occupational exposure in medicine	All occupational exposure, then all exposure of mankind	All exposure of all species
Who/what is being protected	Protection of man alone	Environment assumed protected because man is protected	Demonstration that environment is protected
Known effects of radiation exposure, aims of radiological protection	Prevent deterministic effects...	... and avoid stochastic effects...	... and recognise the possibility of non-targeted effects
The ethical basis of protection	'Respect for life' virtue ethics	Focus on utilitarian ethics	Increasing emphasis on deontological ethics
Protection methods	Advice on practical protection methods	Application of dose limits, then application of optimisation	Optimisation of protection under dose and risk constraints

dominated the use of radiation entirely. Doses to staff were, by today's standards, extremely high, and it was assumed that there was a safe threshold below which radiation would cause no harm. The advent of accelerators and, soon thereafter, of nuclear reactors meant that suddenly there was an abundance of different radionuclides which were used for all sorts of purposes in many different professions. Consequently, the recommendations of the early 1950s were aimed at all occupational uses of radiation. The dawning realisation, during the mid- and late 1950s, that ionising radiation is a genotoxic agent meant that a linear, non-threshold model of radiation – dose response – had to be adopted. As a result, radiological protection had to take account of public exposures and medical exposures of patients, as well as occupational exposures.

3.2.2. Scope of protection

(102) Initially, radiological protection only aimed to protect human beings. *Publication 26* (ICRP, 1977) suggested that 'the level of safety required for the protection of all human individuals is thought likely to be adequate to protect other species, although not necessarily individual members of those species. The Commission therefore believes that if man is adequately protected then other living things are also likely to be sufficiently protected'. Thus, protection of other species was seen as a perk rather than an actual aim. *Publication 60* (ICRP, 1991) re-iterated much the same position, and also clarified that any protection of non-human species was regarded as important only insofar as it may have affected mankind through environmental transfer: 'The Commission believes that the standard of environmental control needed to protect man to the degree currently thought desirable will ensure that other species are not put at risk. Occasionally, individual members of non-human species might be harmed, but not to the extent of endangering whole species

or creating imbalance between species. At the present time, the Commission concerns itself with mankind's environment only with regard to the transfer of radionuclides through the environment, since this directly affects the radiological protection of man'.

(103) The Commission started to think about protection of non-human species in their own right in *Publication 91* (ICRP, 2003) and the 2007 Recommendations (ICRP, 2007). The reason behind this change is not any serious concern about existing radiation hazards. It is rather a matter of filling a conceptual gap by providing scientific evidence, rather than just assumptions, to show that other species are adequately protected.

3.2.3. Known effects of ionising radiation and corresponding protection aims

(104) Anecdotal evidence of radiation-induced cancer was available within less than 10 years of the discovery of ionising radiation, and the mutagenic effect of such radiation had been studied in some detail in experimental organisms before 1930. Nevertheless, before the Second World War, only deterministic effects of radiation were considered in radiological protection, and the aim was to prevent such effects.

(105) During the 1950s, attention to the potential for radiation-induced stochastic effects increased, and in *Publication 9* (ICRP, 1966), this was taken into account in the description of the aims of radiological protection: 'The objectives of radiation protection are to prevent acute radiation effects, and to limit the risks of late effects to an acceptable level'.

(106) More recently, the discovery of non-targeted effects, which cannot easily be classified as either deterministic or stochastic, has complicated the picture. However, the 2007 Recommendations of ICRP state that 'induced genomic instability and bystander signalling ... may influence radiation cancer risk ..., but that current uncertainties on the mechanisms and tumorigenic consequences of the above processes are too great for the development of practical judgements'. Furthermore, 'The Commission also notes that since the estimation of nominal cancer risk coefficients is based upon direct human epidemiological data, any contribution from these biological mechanisms would be included in that estimate'.

3.2.4. The ethical basis of radiological protection

(107) Before the Second World War, the ethical basis of radiological protection was not formally discussed. Its sole aim was to prevent deterministic harm to individual human beings, and this can be seen simply as an example of 'virtue ethics'.

(108) With increasing weight being given to optimisation in the 1960s and 1970s, the recommendations of ICRP were largely based on utilitarian consequence ethics, emphasising what is best for society. The recommendations from ICRP that have been made in the last 10 years have emphasised controls on the maximum dose or risk to the individual. There has been a corresponding reduction in the emphasis on collective dose and cost–benefit analysis. Overall, this reflects a shift of emphasis of the ethical position, paying less attention to utilitarian values. Instead, the

Commission has now increased its emphasis on deontological duty ethics, emphasising what is best for the individual.

(109) Inevitably, radiological protection (and indeed any form of regulation or protection against some noxious agent) will require a balancing between these two ethical principles. No practical protection work can be based on an absolute application of one principle alone; however, one can give more emphasis to one of the principles without entirely discarding the other (Hansson, 2007). This is the development seen from *Publication 26* to *Publication 103*.

3.2.5. Protection methods

(110) The Commission's early recommendations paid considerable attention to the practical aspects of shielding, working methods, and so forth. When numerical advice was introduced, it was in terms of dose limits (or, at least, limits on exposure rates). With *Publication 26* (ICRP, 1977), the emphasis was shifted towards optimisation. *Publication 60* (ICRP, 1991) introduced the concept of dose and risk constraints, with the two purposes of increasing equity (i.e. more emphasis on deontological ethics) and providing a practical tool for the control of multiple sources. However, the level of detail concerning constraints in *Publication 60* was insufficient. This was remedied in *Publication 103* (ICRP, 2007); the purpose and use of constraints is discussed in detail, and hopefully this valuable tool will now be utilised to the full in practical radiological protection.

3.3. References

Hansson, S.O., 2007. Ethics and radiation protection. J. Radiol. Prot. 27, 147–156.

IAEA, 1996. International Basic Safety Standards for Protection Against Ionizing Radiation and for the Safety of Radioactive Sources. IAEA Safety Series 115. International Atomic Energy Agency, Vienna.

ICRP, 1966. Recommendations of the International Commission on Radiological Protection. ICRP Publication 9. Pergamon Press, Oxford.

ICRP, 1977. Recommendations of the International Commission on Radiological Protection. ICRP Publication 26. Ann. ICRP 1(3).

ICRP, 1991. 1990 Recommendations of the International Commission on Radiological Protection. ICRP Publication 60. Ann. ICRP 21(1–3).

ICRP, 2003. A framework for assessing the impact of ionising radiation on non-human species. ICRP Publication 91. Ann. ICRP 33(3).

ICRP, 2007. The 2007 Recommendations of the International Commission on Radiological Protection. ICRP Publication 103. Ann. ICRP 37(2–4).

UNSCEAR, 2006. Effects of Ionizing Radiation. UNSCEAR 2006 Report. Volume 1. Report to the General Assembly, with Scientific Annexes A and B. UN Sales Publication 08.IX.6. United Nations, Vienna.

Main Commission Chairs, Scientific Secretaries, and some other IXRPC/ ICRP personalities

Rolf Sievert, 1923 (Chair 1928-31, 1956-62)

George Kaye (Scientific Secretary 1928)

René Ledoux-Lebard (Chair 1931-37)

Sir Ernest Rock Carling (Chair 1950-56)

Rolf Sievert 1965 (Chair 1928-31, 1956-62)

Sir Edward Pochin (Chair 1962-69)

Gordon Stewart (Chair 1969-77)

Dan Beninson (Chair 1985-93)

Five Scientific Secretaries in 1975: From left, Lauriston Taylor (1934, 1937, 1947-50; also Chair, 1937-50), Eric Smith (1956; also Committee 3 Chair 1962-65), Walter Binks (1950-55), David Sowby (1962-85); Bo Lindell (1957-62; also Committee 3 Chair, 1965-77 and Chair, 1977-85)

Mike Thorne (Scientific Secretary 1985-87)

Hylton Smith (Scientific Secretary 1987-97)

From left: Roger Clarke (Committee 4 Chair 1989-93, Chair 1993-2005), Jack Valentin (Scientific Secretary 1997-2008), Lars-Erik Holm (Chair 2005-2009)

Claire Cousins (Chair 2009-)

Chris Clement (Scientific Secretary 2009-)

Giaocchino Failla (Sub-Committee I/ Committee I Chair 1950-59)

John Loutit (Committee I/ Committee 1 Chair 1959-65)

Warren Sinclair (Committee 1 Chair 1985-2001)

Karl Z. Morgan (Sub-Committee II/ Committee II/ Committee 2 Chair, 1950-73)

Left front: Arthur Upton (Committee 1 Chair 1973-81), left back: Dan Beninson (Committee 1 Chair 1981-85, Chair 1985-93, Committee 4 Chair 1993-97), right: Henri Jammet (Committee 4 Chair 1962-85, Committee 3 Chair 1993-96)

From left: Fred Mettler (Committee 3 Chair, 1996-2005), Hiro Matsudaira (Main Commission member, 1993-2001), Alex Kaul (Committee 2 Chair, 1993-2001)

Charlie Meinhold (Committee 3 Chair 1977-85, Committee 2 Chair 1985-93)

Julian Liniecki (Committee 3 Chair 1985-93)

Val Mayneord (Sub-Committee IV/ Committee IV Chair 1950-56)

John Dunster (Committee 4 Chair 1985-89)

Wolfgang Jacobi (Main Commission member 1977-97)

Li Deping (Main Commission member 1985-97)

From left: Main Commission members Nataliya Shandala (2005-) and Leonid Ilyin (1993-2001)

Annals of the ICRP

Published on behalf of the International Commission on Radiological Protection

Aims and Scope

The International Commission on Radiological Protection (ICRP) is the primary body in protection against ionising radiation. ICRP is a registered charity and is thus an independent non-governmental organisation created by the 1928 International Congress of Radiology to advance for the public benefit the science of radiological protection. The ICRP provides recommendations and guidance on protection against the risks associated with ionishing radiation, from artificial sources widely used in medicine, general industry and nuclear enterprises, and from naturally occurring sources. These reports and recommendations are published four times each year on behalf of the ICRP as the journal *Annals of the ICRP*. Each issue provides in-depth coverage of a specific subject area.

Subscribers to the journal receive each new report as soon as it appears so that they are kept up to date on the latest developments in this important field. While many subscribers prefer to acquire a complete set of ICRP reports and recommendations, single issues of the journal are also available separately for those individuals and organizations needing a single report covering their own field of interest. Please order through your bookseller, subscription agent, or direct from the publisher.

ICRP is composed of a Main Commission and five standing Committees on: radiation effects, doses from radiation exposure, protection in medicine, the application of ICRP recommendations, and protection of the environment, all served by a small Scientific Secretariat. The Main Commission consists of twelve members and a Chair. Committees typically comprise 15–20 members. Biologists and medical doctors dominate the current membership; physicists are also well represented.

ICRP uses Working Parties to develop ideas and Task Groups to prepare its reports. A Task Group is usually chaired by an ICRP Committee member and usually contains a majority of specialists from outside ICRP. Thus, ICRP is an independent international network of specialists in various fields of radiological protection. At any one time, about one hundred eminent scientists are actively involved in the work of ICRP. The Task Groups are assigned the responsibility for drafting documents on various subjects, which are reviewed and finally approved by the Main Commission. These documents are then published as the *Annals of the ICRP*.

This report was drafted by the following Task Group

Full members

W. Weiss (Chairman)
J. Fairobent
M. Morrey
O. Pavlovsky
D. Queniart

Corresponding members

E. Buglova
T. Lazo
I. Robinson